ENVIRONMENTAL SCIENCE, ENGINEERING AND TECHNOLOGY

MUNICIPAL SOLID WASTE

Environmental Science, Engineering and Technology

Additional books in this series can be found on Nova's website under the Series tab.

Additional E-books in this series can be found on Nova's website under the E-book tab.

ENVIRONMENTAL SCIENCE, ENGINEERING AND TECHNOLOGY

MUNICIPAL SOLID WASTE

GIANNI KONNO
AND
CLAUDIO MACHADO
EDITORS

Nova Science Publishers, Inc.
New York

For permission to use material from this book please contact us:
Telephone 631-231-7269; Fax 631-231-8175
Web Site: http://www.novapublishers.com

Additional color graphics may be available in the e-book version of this book.

Library of Congress Cataloging-in-Publication Data

Municipal solid waste / editors, Gianni Konno and Claudio Machado.
p. cm.
Includes index.
ISBN 978-1-62100-257-4 (hbk.)
1. Refuse and refuse disposal--Developing countries. 2. Refuse and refuse disposal--Environmental aspects. I. Konno, Gianni. II. Machado, Claudio.
TD790.M86 2011
628.4'4091724--dc23

2011032384

Published by Nova Science Publishers, Inc. † New York

CONTENTS

PREFACE

Municipal solid waste (MSW) is a commodity resulting from human endeavor and will always exist in the world's cities. It comprises of material originating from households, commercial and medical establishments, construction and demolition activities, street cleaning, and park maintenance. In this book, the authors present topical research in the study of municipal solid waste, including its management in developing countries; the synchronization of solid waste management in urban centers with integrated nutrient management in Indian agriculture; NO and N2O emissions during fluidised bed combustion of lignite and the use of industrial waste and municipality solid waste as an aggregate, filler or fiber in cement mortar and concrete.

Chapter 1 - A general strategy is proposed for establishing and reaching targets of municipal waste management in developing countries. Strategic points of progress are plotted on the time scale established by the 2002 World Summit on Sustainable Development. The turning point occurs when complete collection service is provided. Two twelve-year intervals lead from the initial conditions in any city up to the turning point, and another two intervals trace the progress from the turning point to a zero waste situation. The amount of effort required to satisfy all targets is a function of the initial collection coverage and the demographic expansion, and is quantified in this study. As result, the targeting plot is presented, which in essence can be used with the appropriate local numbers by any municipality of the developing world. The study terminates with the presentation and discussion of key tactical parameters to support the basic strategy. The targeting plot may be returned as a guide for elaborating the directives of future world summits on sustainable development, thus closing the loop.

Chapter 2 - Organic matter status of cultivated lands in most of the agroecological regions of India is low due to rapid mineralization loss of C input into the soil from different sources. This, in conjunction with low fertilizer input, has resulted in continuous decline in agricultural productivity. Long-term fertilizer experiments as well as manurial trials in various parts of the country have conclusively shown that integrated nutrient management (INM) involving various types of organic manure can sustain or augment agricultural productivity. Rural compost prepared mainly from cattle shed litters and agricultural wastes remained major source organic matter in the INM in Indian agri-horticulture. However, availability of rural compost is very low (on average, little over 2 t/ha) and is declining continuously due to decreased dependency on animals in crop husbandry.

On the other hand, India generates about 70 million tonne (MT) of municipal solid wastes annually, which has the potential of producing about 9 MT compost (MSWC). This has the huge potential of supplying important plant nutrients N (55 thousand tonne / year), P (14 thousand tonne / year), K (40 thousand tonne / year) and Zn (2.3 thousand tonne / year) to cultivated land besides providing organic matter for improvement of physical and biological environment. A survey in 29 cities has indicated that quality of MSWC in respect of fertilizing potential and heavy metals contents varied widely with the processes adopted for manufacturing. In general, fertilizing potential of the MSWC in majority of the cities is very low, resulting in their poor marketing. Such a situation has ultimately caused poor volume (presently about 9%) of recycling of solid wastes into compost. This review intends to deal with three issues; namely, demand of organic matter for INM in agriculture, recycling of MSW towards compost manufacturing, and protection of soil resources from heavy metal pollution. A new system of classification for MSWC into different marketable classes (A, B, C, and D) and restricted use classes (RU-1, RU-2, and RU-3) has been proposed on the basis of their fertilizing potential and heavy metal pollution potential. India has diversified cropping system which includes foodgrain crops, non-food crops, and horticultural crops covering about 71%, 23% and 6% of total gross cropped area. Potential use for each of the classes of MSWC in INM of different agri-horticulture systems has been discussed keeping in view of nutrient requirement as well as probability of contamination of environment. This classification system is expected to maximize recycling of MSW, reduce the burden of land-fill sites, reduce the gap between demand and supply of organic matter for INM, improve soil physical environment and reduce environmental pollution by mitigating emission of methane (a green house gas) from landfill sites. Ways of improving present system of solid waste management for producing good quality compost have also been suggested.

Chapter 3 - Municipal solid wastes are accepted less harmful than the other kind of wastes in terms of environmental and public health depending on the process of their occurrence. The wastes determined as municipal wastes involve malfunctioned forms of products such as organic substance, plastic, textile, paper, wood, metal and glass. The proportion of these substances is being changing in accordance with the cultural structure and life standards of societies. In the world, there are various approaches for controlling municipal solid wastes. Yet the methods used for the solution of the problem have been insufficient in their own. The gretest problem arising from solid waste is primarily the increasing amount of these wastes with each passing day. The focus of the studies performed recently on this issue is to immobilize the amount of waste or at least slow down the rate of increase. However, nowadays, it is clear that the situation will conflict with the economic policies supporting the consumption growth and needs long-termed studies.

For waste management the necessity of building on the acceptance of waste raw materials by taking into account the durability of needs and limitedness of natural sources is obvious. Generally, a successful waste management should keep various eliminating methods together and be supported with laws in order to get ahead with the methods applied. The significant point in here, and particularly within the scope of waste-raw material relation, is that any society being in the position of producer should be a part of the process of waste management and be encouraged to increase efficiency of recycling. As well as increasing the success of recycling processes by allocating responsibility to its producer with regard to recognizing and separating its waste, the control of hazardous wastes arising from domestic use also gains importance. Thus, enabling municipal wastes to be in less harmful position will lead to progress in the context of the environment protection in a short time.

Chapter 4 - An experimental investigation on the co-firing characteristics of Thai lignite and municipal solid waste (MSW), with emphasis on the emissions of major nitrogenous-species, has been carried out in a bubbling fluidised bed. The lignite used was of 3% sulphur and the MSW was simulated using materials representing the main compositions of high moisture MSW, that is, a mixture of vegetable, paper, bark, plastics and crushed brick. The effects of co-combustion on temperature distributions, combustion zone and carbon combustion efficiency were investigated, and their implications on the influence of the mass fraction of MSW in the MSW/lignite mixture on the emissions of NO, N_2O, NH_3 and HCN, measured using an FTIR gas analyzer, were discussed. Tests were run with MSW mass fraction varied between 0 and 40% and excess air between 40 and 100%. The results show that the addition of MSW results in a drop in bed temperature and a slight decrease in carbon combustion efficiency, due primarily to the high moisture content of MSW. The high volatile content of the MSW promotes combustion in the freeboard, leaving less fuel particles for in-bed combustion. This phenomenon is most obvious in the case of 20% MSW co-combustion, where the synergistic effect of high volatility on the reactivity of the mixture is believed to be the highest. The emission of NO generally increases with increasing mass fraction of MSW due mainly to the increase in the release of NH_3 from the MSW and a decrease in catalytic char particles for NO reduction. NO appears to be minimised at 20% MSW. N_2O decreases slightly to a minimum at 20% MSW due to temperature decrease, but increases as excess air increases further because of increasing presence of CH_4 and NO. The emission of NH_3 is much more significant than HCN. Overall, the total fuel-nitrogen converted as NO, N_2O, HCN and NH_3 increases with MSW mass fraction, except at 20%.

In addition, the comparison of 35% waste moisture and 60% waste moisture were preformed. The results show that the bed temperature in the case of 35%-moisture is higher than for in case of 60% due to the difference of physical properties of the fuel. The combustion efficiency for the case of 35% waste moisture is higher than that for 60% moisture due to higher bed temperature at lower waste moisture. The synergistic effect of the co-combustion of lignite with the drier MSW reduces the emission of CO leading to increase in combustion efficiency. The fuel mixture with low-moisture in waste gives higher fuel-N conversion to NO whereas the fuel-N conversion to N_2O is higher for higher moisture waste, particularly at high excess air.

Chapter 5 - With increasing the greenhouse gases (GHGs) gas emissions into the atmosphere and the possible consequence for the global climate change, there has been a growing interest in develop efficacy management strategies of reducing anthropogenic GHGs emissions.

Municipal solid waste (MSW) management plays an important role in the GHGs emission. The contribution of the landfilled MSW to GHGs it was reported to be of 5% of the total GHGs emitted and of the 13% of the methane from anthropogenic fonts (USEPA, 2003; IPCC, 2006).

The compute of the total GHG emissions is determined quantifying the GHG gases from all human activity as credits and debits. Debits are the GHG emitted in the atmosphere; on the other hand, the credits are the storage on C in stable forms (C-based systems) (IPCC, 2006; Brown et al., 2008).

Because the production of CH_4 *vs* CO_2 during MSW degradation depends by aerobic *vs* anaerobic conditions, the possibility to manage MSW organic fraction by aerobic biological process, i.e. composting and mechanical biological treatment, avoiding methane production in landfill, reduce GHGs emission due to MSW management.

The potential reduction of the methane emission and the possibility to store C in landfill and soil under recalcitrant forms, is extensively discussed in this chapter, reporting both literature and new data.

Chapter 6 - Management and treatment of industrial solid waste and municipal waste has been gaining importance. Waste material can be utilized as fuel in cement kilns, as raw material for cement clinker, as mineral addition to cement, or as granular material in cement mortar and concrete. Depending on the properties of the waste, it can be used in the cement and concrete industry without treatment or after treatment. Although vast amounts of waste material can be consumed in the production of cement clinker and blended cement, that consumption can be increased substantially if it is used as aggregate in cement mortar and concrete. Using waste material in this way can solve problems of shortage of aggregate in construction sites and reduce environmental problems related to aggregate mining and waste disposal.

For effective use of waste in the construction industry, a combined chemical, civil, material and environmental engineering approach is necessary. Careful characterization of waste is necessary before it is used in construction. Treatment of some waste materials is necessary to improve their quality before being incorporated into concrete. Environmental assessment of waste and concrete containing waste is also necessary for proper evaluation of waste materials as a constituent in concrete.

Slump, bulk density, unit weight, and air content; mechanical properties such as strength and elasticity modulus, and durability properties such as water absorption, shrinkage, carbonation and chloride resistance of fresh and hardened concrete containing waste materials, are normally evaluated. Some waste materials can improve certain concrete properties such as fire behaviour, thermal insulation and cracking, and therefore some reports are available on these topics too. But the durability performance of concrete containing waste materials has not been evaluated to the same extent and therefore significant research is necessary on this topic. As different waste materials can change the microstructure of mortar and concrete and the hydration chemistry of cement, more work is necessary to understand the aggregate-cement interaction in concrete and to develop more robust waste aggregate-based concrete technology.

In this chapter, the properties of plastic, municipal solid waste incineration ashes, coal bottom ash, tire rubber waste, different industrial slag and waste materials from miscellaneous industries when used as aggregate or filler or fibre in cement mortar and concrete are discussed. For proper presentation of the results the chapter is divided into subsections for each type of waste material. Finally, a concluding comment along with the scope of future research is presented.

In: Municipal Solid Waste
Editors: G. Konno and C. Machado

ISBN 978-1-62100-257-4

Chapter 1

PLOTTING TARGETS FOR A ZERO WASTE STRATEGY ON THE WORLD SUMMIT TIME SCALE

Manfred Fehr *

Institute of Geography, Federal University at Uberlândia
MG, Brazil

ABSTRACT

A general strategy is proposed for establishing and reaching targets of municipal waste management in developing countries. Strategic points of progress are plotted on the time scale established by the 2002 World Summit on Sustainable Development. The turning point occurs when complete collection service is provided. Two twelve-year intervals lead from the initial conditions in any city up to the turning point, and another two intervals trace the progress from the turning point to a zero waste situation. The amount of effort required to satisfy all targets is a function of the initial collection coverage and the demographic expansion, and is quantified in this study. As result, the targeting plot is presented, which in essence can be used with the appropriate local numbers by any municipality of the developing world. The study terminates with the presentation and discussion of key tactical parameters to support the basic strategy. The targeting plot may be returned as a guide for elaborating the directives of future world summits on sustainable development, thus closing the loop.

INTRODUCTION

Municipal solid waste (MSW) is a commodity resulting from human endeavor and will always exist in the world's cities. It comprises refuse originating from households, commercial and medical establishments, construction and demolition activities, street cleaning and park maintenance. The rate of production is commonly reported on a per-person-per-day basis, which implies that it will grow together with the population. The

* http://www.manfredfehr.com.br, prosec22@yahoo.com.

administration of waste movement from its point of generation to its final destination has to obey local parameters such as waste composition and existing reverse logistics. Separate schemes are normally required for the various types of waste, and no universally valid procedure is available.

Household waste (HW) originates from normal daily activities in municipal residences. Generation rate and composition depend on geographical location as well as on social stratification of communities. HW is a quite heterogeneous commodity that comprises biodegradable as well as inert material in various proportions and states of mixture. The relation between biodegradable and inert material varies all the way from 70/30 in the developing world to 30/70 in developed countries (Fehr 2002). This has implications in collection as well as in treatment strategies. The infrastructure for separately collecting and processing biodegradable waste e.g. is the main cost item and management bottleneck in developing countries and therefore still is practically non-existent. Due to the much smaller size of this component in developed countries, mechanical-biological pretreatment (MBP) facilities satisfactorily attend to it.

Commercial refuse varies from almost pure biodegradable material in the establishments catering to the food sector to almost pure inert material in the case of stationery, clothing and furniture retailers. Both household and commercial wastes form the topic of this chapter, as they are the main visible items of urban sanitation, are the direct responsibility of the municipal administration and as such have been targeted by world summit directives.

Construction debris and health care refuse are special categories of municipal waste the responsibility of which is usually assigned to the producers by specific national or municipal legislation. Material resulting from street cleaning and park maintenance, which consists mainly of soil, wood, leaves and some inert items can readily be sorted and integrated into landscaping projects.

In the not too distant past, waste was conceived of as a nuisance that could be removed by collection followed by tipping at a site not visible to the population: a landfill at best.

At the turn of the century, the United Nations Organization introduced quantitative targets into global and local sanitation policies. The Millennium Goals of 2000 included environmental sustainability generally (United Nations 2000) and the documents originating from the Johannesburg Summit of 2002 referred to the advancement of sanitation services specifically (United Nations 2002). The precise target of the Johannesburg Summit was and still is to reduce by one half the proportions of people excluded from sanitary services within a timeframe of 12 years, from 2003 to 2015.

The first challenge for municipal administrators refers to the correct interpretation of the directive. What exactly is meant by providing sanitary services, or more specifically MSW management services to a given proportion of the population? This is a management problem. Sanitation services are usually provided to residences, not to individuals. So the starting diagnosis, the yearly progress requirements and the target all have to be spelled out in terms of the number of residential units. This procedure produces numbers quite different from the population count. From the point of view of a residential unit, the service is considered complete if the trash is taken away regularly. From the point of view of municipal administrations, the problem is not resolved without finding and using an adequate destination for the trash. This is a policy problem. The fact remains that the Summit directive implied collection service only. Nothing was specified on destination, so the landfills not only maintained their status but kept growing.

The delegates to the Summit probably did not anticipate the fundamental split of interest their ordinance would cause within the sanitary engineering community. Geographical considerations play a vital role in measuring the consequences of the reduction-by-half demand issued at the Summit. For Northern countries, which already had satisfactory universal sanitation infrastructure in place in 2002, the directive had no bearing at all. Southern countries, on the other hand, found themselves directly addressed as they evidenced considerable gaps between supply and demand of sanitary services. National as much as municipal administrations are still struggling to meet the target set forth for 2015, and management models are in high demand.

The Summit directive was pragmatic and concise. No details of any kind were addressed. Consequently, municipal administrations in Southern countries face crucial decisions of procedure and infrastructure deployment.

Furthermore, the concept of a landfill as waste receiver has been challenged in recent decades for reasons of municipal sustainability, space availability and collateral effects on soil, water and air. World summits have addressed the topic of waste within the general subject of sanitation, and have set targets for the gradual expansion of collection services to urban residents. This development has resulted in enormous management challenges for municipal administrations. Within a period of only three decades, waste management targets over and above the summit directives have moved from 100% landfilling to 0% landfilling, and administrative recipes are in high demand. This author's research has identified source separation as the main driver of diversion of household and commercial waste from landfills or incinerators in emerging economy contexts. The reverse logistics infra-structure is largely informal, but it exists and diverts source separated material from landfills without cost to the public budget.

According to the policy decision, various management models may arise with their corresponding challenges for the sanitary engineering community, and these challenges may be technical as much as social or managerial.

In the case of a management model that stipulates bulk collection and tipping at a landfill, what is the prospect for engineering firms? There is high demand for construction and operation of landfills, and for manufacturing and maintaining collection vehicles; in short, a bright business future with doubtful sustainability. In the case of a management model that stipulates "zero waste" as final target, the engineering challenge takes on very different forms. There is no landfill and no bulk collection. The challenge is pushed backward through the logistics chain and decentralized. Recycling considerations begin long before the material ever reaches the waste baskets.

The two extreme cases mentioned above illustrate, respectively, the traditional technical demands and the new social demands engineering firms face in the waste business. From the vantage point of emerging economy contexts, the present study contributes experimental results as well as strategic instruments to the task of passing from the first to the second extreme within a timeframe based on the Summit directive.

Elaborating on the ideas and guidelines found in the literature and on the author's experimental work, the present chapter proposes to overcome the apparent vacuum in the local decision making process related to strategy. It is argued that when working toward a landfill diversion or zero waste goal, strategy precedes tactics. Consequently, a generally applicable strategy is developed for setting landfill diversion targets in any local context that may easily be quantified and measured. The study adopts the time scale set forth by world summit directives, i.e. twelve-year intervals. This means that starting from any local situation with respect to municipal waste, four basic targets are defined. They are twelve years apart and can lead to a zero waste situation within 48 years. The first target is adopted directly from the last Summit directive and leads to reduction by one half of the proportion of residences without waste collection. The second target builds on the first and predicts the most likely directive to be emitted by the next summit, namely achieve universal waste collection. The third target is based on experiment. It sets the threshold for recycling by private initiative, which may differ from one local community to another. The fourth target, fatally, is total landfill diversion or zero waste or "darn close to it". It requires direct intervention of the public administration to complete the cycle in 48 years when the last landfill will be closed.

Once the strategic targets are established, the tactics of operation and procedure may be developed, and ideas are readily available in the literature. They refer to dominating the reverse logistics for biodegradable material and to creating recycling options for material the private sector has been unable to absorb. An important consideration refers to possible backfire from contracts with municipal concessionaires. As long as landfills and incinerators exist, their operators naturally oppose recycling because it curbs their profits. This detail needs to be addressed by appropriate contract formulation. The proposed scheme is meant for emerging economies and other developing countries. For developed countries, the targets of the first two 12-year terms are likely to be innocuous because they were reached long ago, so the total timeframe would be much shorter. Any municipality of the developing world can use the proposed strategy with minor adaptations where required, and future world summits can build their directives on it.

Literature on Municipal Waste Management Tactics

Municipal waste management has been treated exhaustively in the scientific literature. A survey shows, however, that this treatment has largely been concerned with tactics such as recycling technologies, collection practices, reverse logistics schemes and economics of procedure. Governance strategies such as precise targeting for progressive landfill diversion of waste have not been addressed, although they are the single most urgent need of municipal administrators. The United Nations Environmental Program (UNEP 2010) and the Zero Waste International Alliance (ZWIA 2010) have issued general definitions and guidelines for moving a community toward a zero waste situation, without specifying precise targets and deadlines. The training manual of the United Nations Environmental Program provides mathematical theory on objectives and targets, but does not show how to establish them in a local context. The Zero Waste International Alliance has gone much further. It has issued a definition of zero waste as well as the corresponding business and community principles. The Alliance encourages communities to pursue the goal of diverting 90% of waste from landfills or incinerators within 15 years of adopting the plan, but does not reveal how it reached those numbers. World summits on sustainable development have contributed directives for local targeting such as reducing by one half within twelve years the proportion of people not served by basic sanitation. This requirement applies to municipal waste, but obviously does not enter into details of procedure (United Nations 2002).

The International Environmental Technology Center of UNEP (UNEP 2010) has published a training manual for integrated solid waste management (ISWM) consisting of four volumes. Volume 1 is concerned with procedure to define waste streams, to collect and analyze samples and to present the data. Volume 2 dwells on the identification of current status of waste management in a community, including policies, technology and law enforcement. The subject of targets is addressed in Volume 3, but in a semantic way. A target is defined by its dictionary entry as "the situation one intends to achieve and that is influenced by action; it must be measurable and must have a timetable". Finally, Volume 4 provides the semantics of a management plan that "needs to establish goals, targets, objectives, strategies and tactics". The manual is meant for students of waste management, so it does not deal with real world numbers. The present chapter complements the UNEP manual by introducing a pragmatic approach to assigning definite numbers to targets and deadlines in a local context.

The Zero Waste International Alliance defines the zero waste goal as "to eliminate all discharges to land, water or air that are a threat to planetary health". It encourages the industrial sector to divert more than 90% of its waste from landfills or incinerators, and residential communities to establish benchmarks and timelines to reach 90% diversion within 15 years of adopting a plan. The contribution of the present chapter to this line of reasoning is concerned with proposing to local administrations absolute numbers on the time scale. The general 90%-in-15-years diversion target is replaced by an experimental progressive diversion plot over 48 years suited for developing nations.

Domestic Waste in Emerging Economies

The scientific literature on domestic waste analysis proper in different countries tends to be repetitive. More than ten years ago, this author's team analyzed several tons of mixed domestic waste in a typical urban context in Brazil and produced an average result from various geographical and social situations. The data are reproduced in column 1 of Table 1.

Table 1. Unsorted domestic waste compositions reported from various developing countries, in percent

	Brazil	Nepal	Nigeria	Mexico	India	China
biodegradable material	72	71	72	81	71	72
plastics	11	12	11	5	9	11
paper and cardboard	7	8	3	7	9	7
glass	3		4	4	5	5
metals	3		2	3	4	1
textiles	2		5		2	2
miscellanea	2	9	3			2

Sources: Castro 1998; Calçado1998; Fehr and Castro 1999; Fehr and Calçado 2001; Fehr et al. 2010; Dangi et al. 2010; Nabegu 2010; Benítez et al. 2003; Chatterjee 2009; Qu et al. 2009.

Many similar analyses have since been reported from other parts of the developing world with almost identical results. They are compared in Table 1. A study from Nepal shows the following composition: organics 71%, plastics 12%, paper 8%, others 9% (Dangi et al 2010). This kind of perfect agreement reported from the other side of the Earth cannot be a coincidence. A study from Mexico states the following composition: recyclable organics 57.73%, plastics 3.66%, paper 4.93%, glass 2.8%, metals 1.91%, non-recyclables 28.97% (Benítez et al 2003). The analysis contains a considerable quantity of non-recyclables, which is not defined and probably consists of mixed material that was not separated during the research. If this mixed portion were distributed proportionately among the other items, the composition would be: biodegradables 81%, plastics 5%, paper 7%, glass 4%, metals 3%, as shown in Table 1. Another study originates from Nigeria. It relates the following composition: biodegradables 72%, plastics 11%, paper 3%, glass 4%, metals 2%, textiles 5% miscellanea 3% (Nabegu 2010). A recent analysis originating from India lists as components: biodegradables 62.8%, plastics 8.2%, paper 7.9%, glass 4.1%, metals 3.8%, textiles 1.5%, miscellanea 11.7% (Chatterjee 2009). Just as was the case in Mexico, the large portion of miscellanea indicates that the separation was not carried to conclusion. Again, if this mixed portion were distributed proportionately among the other items, the composition would be: biodegradables 71%, plastics 9%, paper 9%, glass 5%, metals 4%, textiles 2% as shown in Table 1. A report available from China states this composition: biodegradables 69.67%, plastics 10.79%, paper 6.52%, glass 5.01%, metals 0.79%, textiles 2.31%, miscellanea 2.03% (Qu et al 2009). This analysis does not close at 100%. When corrected, it becomes: biodegradables 72%, plastics 11%, paper 7%, glass 5%, metals 1%, textiles 2%, miscellanea 2% as stated in Table 1.

Those reports all come from different continents and are practically identical. It can be concluded that the differences in unsorted domestic waste composition between various parts of the developing world are negligible. Approximately 73% of their household waste is biodegradable, and approximately 27% is biologically inert and consists of plastics, paper, glass, metals and textiles. That fact was recognized by the present author nine years ago (Fehr 2002) and has since been confirmed.

A few additional studies have been conducted in developing countries, but they do not contribute new data to unsorted waste composition. Carmo and Oliveira (2010) assess the prospect of scavenger cooperatives in the recycling business in Brazil. Atasoylu et al (2007) relate that in Turkey only 6.4% of garbage is recyclable. Aini et al (2002) state that in Malaysia only 5% of all domestic waste is recycled and recommend environmental education as means to improve this number. A similar study from Thailand concludes that improved municipal collection procedures drive scavengers out of work (Sembiring and Nitivattananon 2010). Research from Mexico calls upon government to provide adequate disposal sites for municipal waste (Buenrosto and Bocco 2003).

From the point of view of waste management in general, no further analyses of unsorted domestic waste appear to be required. There is but one fallacy in all those reports. They do not reveal the fact that unsorted waste composition cannot be reproduced by source separation in urban residences, and thus is not a management tool. The present author has experimented extensively with guided source separation, and has arrived at reliable reports on sorted domestic waste composition that are adequate as bases for municipal management models (Fehr et al 2010). What remains to be done in urban settings of the developing world, in order to construct workable waste management schemes, is this: experiment with source separation to obtain reliable composition data of sorted waste, study the reverse logistics facilities and resources available in each setting to stimulate their continued and improved contribution to landfill diversion, and last not least, develop management strategies with precise targets and deadlines to engage municipal administrations over long periods of time. The strategy consideration will be dealt with later in this chapter. The literature on waste management originating from the developed world dwells largely on the tactics of procedure for isolated cases. Once the strategies are decided upon, those tactics may be called in to complete the municipal management schemes. Examples follow.

Contributions to Procedure from the Developed World

The Tactics of Domestic Waste Collection

The options considered are curbside collection at individual residences and collection from voluntary drop-off points distributed within the neighborhood. Tralhão et al (2010) present a mathematical model that determines optimal locations of drop-off points with respect to distance from dwellings and collection cost. Jenkins et al (2003) find that access to curbside recycling has a significant positive effect on the percentage of recycled material. Ando et al (2005) identify a relation between recycling rates and distance to recycling bins. Beatty et al (2007) argue that individual curbside collection actually cannibalizes returns from drop-off centers.

The Tactics of Destination for Collected Waste

Various life cycle analysis studies compare the effect of landfilling and incineration on municipal sustainability. Hong et al (2010) find that landfilling has the highest environmental effect because of methane emissions. They propose methane capture for production of electricity. This argument is somewhat out of date. The European Landfill Directive has outlawed this procedure a decade ago. Morris (2005) presents calculations to show that recycling of biologically inert material such as plastics, paper, glass and metals is superior to landfilling with energy recovery in terms of material life cycles, because it decreases raw material extraction from nature. The counterargument to that theory comes from Bohm et al (2010) who conclude that recycling is more costly than bulk collection and disposal. A sensitivity analysis performed by Eriksson et al (2010) shows how to determine the most important parameters in any waste management model.

The Tactics of Economical Incentives for Source Separation

Yan (2010) reports on experiments with reward schemes in 122 private housing estates in Hong Kong and concludes that economic incentives favorably affect recycling behavior. The egoistic factor of recycling behavior is dealt with by Ewing (2001) who shows that people derive satisfaction from their neighbors' compliments. Unit based pricing for household waste is found to increase the fraction of waste that is recycled (Dijkgraaf et al 2004). Ferrara (2003 and 2005) reports from Canada that a system of legal disposal and recycling facilities and a user fee for collection are sufficient to induce households to optimally allocate their resources. A study by Basili (2006) finds that households as well as enterprises are willing to pay for recycling in order to divert waste from landfills. In a study from Taiwan, Yang and

Innes (2007) conclude that unit pricing of garbage, mandatory recycling and charging for plastic bags all reduce garbage production. Dresner and Ekins (2010) develop a scheme that would avoid additional net cost to producers if charging for waste were legalized in England. Lavee and Khatib (2010) propose to establish a benchmark for economically efficient recycling by identifying and imitating pioneering cities.

The Tactics of Determining the Intervention of Public Administrations

Gallardo et al (2010) analyze different selective collection systems and find that the highest recycling efficiency is obtained with a five-bin scheme that separately collects paper, glass and lightweight packaging from drop-off facilities and organics and mixed waste from curbside bins. Shinkuma (2003) compares refund payment for bottles to unit pricing where the cost of the bottle is included in the price of the beverage, and concludes that the choice depends on the value of the bottle and the transaction cost of returning it. Garcia et al (2005) convert biodegradable waste into animal feed stuff by heating and drying at 65°C for 20 minutes, thus defining a destination for this waste fraction. Guimarães et al (2010) and Chen (2010) propose methods and procedures to measure and report the performance of waste utility companies.

Additional Considerations

All the research reported in the literature and cited here is concerned with tactics. The development of a strategy with precise landfill diversion targets has not been addressed.

Waste composition has only been reported *in natura*, meaning that all the analyses found in the literature refer to unsorted waste prior to any sorting effort at the source. Although this kind of composition report is scientifically correct, it does not contribute anything to solving the practical problem of diverting the waste from the landfill. It is of no use to the municipal administrator to know e.g. that 11% of domestic waste is paper, if there is no way to separate this paper from the mixed lot. The present author has experimentally established the fact that source separation drives reverse logistics, and not vice versa. Consequently, landfill diversion by reverse logistics operators is limited to the portion of waste separated at the source. Source-sorted waste composition, then, is the all important parameter for constructing workable schemes for collection and recycling. This simple truth is completely absent from the literature reports on domestic waste analyses.

A few additional comments on the tactical proposals from the literature are in order. The main fallacy of many studies is the fact that they continue to consider a landfill or an incinerator as essential parts of a municipal waste management scheme. This line of reasoning is no longer valid in the 21st Century, as worldwide intellectual movements such as the Zero Waste International Alliance (ZWIA 2010) and the Ecological Footprint Network (EFN 2010) are actively pursuing the goal of a zero waste society. The argument that it is cheaper to tip the garbage than to recycle it cannot be sustained when the opportunity cost of a landfill is included into the calculation. The proposal of tipping biodegradable waste and capturing the methane for power generation is a highly polemic procedure. Apart from leaving the material cycle open, it pollutes soil and ground water and therefore has been banned in Europe more than ten years ago.

It is this author's conviction that developing nations need to jump over some of the paradigms established by developed countries in the past only to be abandoned later.

Landfills, incinerators, central sorting facilities and voluntary drop-off points are examples. Landfills may at most be considered temporary trouble shooters within a waste management strategy that adopts world summit timetables in quest for a zero waste situation. Central sorting and distributed drop-off facilities have been experimented with in Brazil, but have not produced a return compatible with the investment (Fehr et al 2010).

Methods

The methodological approach to landfill diversion with a zero waste scenario in mind, described hereafter, builds on the omissions detected in the literature. To start with, unsorted and sorted waste compositions at the source are experimentally established and compared. With the knowledge of sorted waste composition, a first diversion threshold may be determined and transformed into a practical progress target on the waste management time scale. As the literature has clearly confirmed the similarity of domestic waste compositions in developing nations, the approach to target plotting to be outlined hereafter is considered generally applicable in those countries, even though the numbers referring to sorted waste composition may differ. Human endeavor and behavior need to be assessed in each setting in order to find the sorted waste composition and fix the corresponding threshold of diversion. The method pursues a zero waste situation, and as such discards the landfill as central facility of a waste management scheme in the long term. As initial target corresponding to the last World Summit directive and considered compulsory for any municipal administration, the reduction-by-one-half of the proportion of people not served by sanitary provisions is applied. The twelve-year time span determined by the Summit is adopted as a benchmark not open to discussion. Accordingly, further twelve-year intervals are established on the time scale for successive progress targets on the road to zero waste. Although the intermediate targets may vary from one municipality to the next, the zero waste situation represents the final point on a 48-year scale. The specific twelve-year targets envisaged by this study are respectively: Reduction by one half of the proportion of residences without access to collection; Complete collection service for the population; Landfill diversion of collected waste corresponding to the sorted waste threshold; Zero waste. This progressive targeting approach is similar to the building block method proposed by Falkner (2010) for future summits on climate change. It develops different elements of governance in an incremental fashion and embeds them into the strategic framework leading to the final ambitious zero waste goal.

The tactics to support this strategy may be extracted from the literature, where an abundance of proposals is available. For the specific case of an emerging economy, object of this research, a number of basic principles related to tactics are established and will be described as starting points for adoption elsewhere.

RESULTS AND DISCUSSION

The Difference in Effort for the First Twelve Years

The last World Summit directive caught many municipal administrations by surprise. In spite of the great variety of local situations, the directive related to waste was uniform: reduce by one half the proportions of people excluded from collection services. In order to develop a strategy, the directive needs to be defined more clearly in local terms. Since no waste destinations were indicated in the directive, it is assumed here that it referred to collection only, and thus admitted the continued increase of landfills. It is also assumed that it referred to residential units and not to individuals, because waste collection service is not provided to individual citizens. Furthermore, municipal collection services are only provided to legally occupied territory. Any invaded urban spaces do not qualify for waste collection.

With those restrictions in mind, the timetable for the first twelve years may be described as follows.

2002. The World Summit emits the directive to halve, by 2015, the proportion of people without access to basic sanitation. In the present study this reduces to solid waste collection from residences.

2003. The city administration produces its diagnosis of present waste management practice and establishes the basic parameters for planning. This includes the number of residences, their demographic expansion rate and present coverage of waste collection. From that information, the administration defines the targets and the annual effort required to comply with the directive.

2004. The twelve-year clock starts ticking.

2015. At the end of the twelve-year period, results are reported on the success or failure of the effort to satisfy the directive.

The basic parameters that determine the size of the challenge are

1. The initial number of residences.
2. The annual demographic expansion of the residential universe.
3. The initial collection coverage in percent of the residential universe.

In the course of quantifying the local effort as function of those parameters, a baseline situation was chosen as follows. Initial number of residences = 140000, annual expansion of residences = 1.1%, initial waste collection coverage = 60% of residences.

The study then considered variants of this baseline situation in order to determine the impact of each parameter on the size of the effort in terms of percent annual increase of collection service required to reach the target.

Table 2 shows the results for variants in the number of residences. Independently from the initial universe, the collection coverage has to increase by 3.55% per year in order to reach the target. This result was really predictable because the coverage parameter is given in percent and not in absolute numbers.

Table 2. Annual increase in collection effort with variable numbers of residences, Constant initial coverage 60% and demographic expansion 1.1%/year

2003				
existing residences	140000	160000	180000	200000
residences served	84000	96000	108000	120000
2015				
existing residences	159640	182446	205252	228057
fraction served	0.8	0.8	0.8	0.8
residences served	127712	145957	164201	182446
annual increase	3.55%	3.55%	3.55%	3.55%

Sample calculations:

$140000 \times 1.011^{12}=159640$ 140000x0.6=84000 159640x0.8=127712

0.6+(1-0.6)/ 2=0.8 127712/84000=1.520 increase in 12 years

$1.520^{1/12}=1.0355$ annual increase 3.55%

Table 3 shows the results for variants in the demographic expansion. In this case, the required effort increases with the expansion ratio.

Table 3. Annual increase in collection effort with variable demographic expansion, Constant initial coverage 60% and initial number of residences 140000

2003				
existing residences	140000	140000	140000	140000
residences served	84000	84000	84000	84000
demographic expansion	0.9%/year	1.1%/year	1.4%/year	2.0%/year
2015				
existing residences	155891	159640	165418	177554
fraction served	0.8	0.8	0.8	0.8
residences served	124713	127712	132335	142043
annual increase	3.35%	3.55%	3.86%	4.47%

Sample calculations:

$140000 \times 1.009^{12}=155891$ 140000x0.6=84000 155891x0.8=124713

0.6+(1-0.6)/2=0.8 124713/84000=1.485 increase in 12 years

$1.485^{1/12}=1.0335$ annual increase 3.35%

Table 4. Annual increase in collection effort with variable initial coverage. Constant demographic expansion 1.1%/year and initial number of residences 140000

2003				
existing residences	140000	140000	140000	140000
fraction served	0.4	0.6	0.8	0.9
residences served	56000	84000	112000	126000
2015				
existing residences	159640	159640	159640	159640
fraction served	0.7	0.8	0.9	0.95
residences served	111748	127712	143676	151658
annual increase	5.92%	3.55%	2.10%	1.60%

Sample calculations:

$140000 \times 1.011^{12} = 159640$ $140000 \times 0.6 = 84000$ $159640 \times 0.8 = 127712$

$0.6 + (1 - 0.6)/2 = 0.8$ $127712/84000 = 1.520$ increase in 12 years

$1.520^{1/12} = 1.0355$ annual increase 3.55%

Table 5. Comparison of required collection efforts for cities with different starting points. Constant initial number of residences 140000. Required annual increase in collection services in percent

demographic expansion ratio initial collection coverage	0.9%/year	1.1%/year	1.4%/year	2.0%/year
90%	1.35	1.60	1.85	2.46
80%	1.90	2.10	2.40	3.00
60%	3.35	3.55	3.86	4.47
40%	5.71	5.92	6.24	6.87

Table 4 shows the results for variants in the initial collection coverage. The annual effort has a reverse relation to the initial coverage.

Finally, the results for the variants are compared in Table 5 where the tendency of increasing effort proceeds diagonally from the top left to the bottom right. This table exhibits the surprises hidden in the World Summit directive. Depending on the initial situation, different municipalities face quite different demands of service expansion. As an example, a city with initial coverage of 40% and demographic expansion of 2.0% per year faces a coverage expansion requirement five times that of another city with 90% coverage in place and demographic expansion of 0.9% per year.

With those considerations, the first twelve-year period following the last summit has been quantified in terms of collection obligations. The directive was pragmatic and precise. All cities had to reduce by one half the number of residences excluded from collection. This demand was not negotiable. It goes without saying, though, that if any city does not comply, the demand does not go away. The twelve-year period in this case may be moved along the time scale and the target reached behind schedule, but there is no excuse for omission.

Prospects for the Second Term of Twelve Years

In an attempt to predict the directive originating from the next world summit, two options are explored as obvious candidates. In the first option, all municipalities are requested to continue after 2015 with the same collection effort as in the first twelve-year period, until reaching a situation of complete collection service. This will occur at a point in time specific to each city. In the second option, a new twelve-year time frame is arbitrarily established for all municipalities to reach the universal target of complete collection service.

In order to illustrate the two choices, the two cases with extreme efforts from Table 5 are considered. The first municipality deployed an effort of 1.35% annual increase in collection to pass from 90% to 95% collection service in the presence of a yearly demographic expansion of 0.9% between 2003 and 2015. The second municipality deployed an effort of 6.87% annual increase in collection to pass from 40% to 70% collection service in the presence of a yearly demographic expansion of 2.0% between 2003 and 2015. The two "post-2015" options for those extreme cases are presented in Tables 6 and 7.

Table 6. Strategy options after 2015. Option 1 – Maintain collection effort of last 12 years. Constant initial number of residences 140000

City A		City B
0.9%/year	demographic expansion	2.0%/year
90%	initial collection coverage	40%
	year 2003	
140000	initial number of residences	140000
126000	residences served	56000
	year 2015	
155891	existing number of residences	177554
95%	collection coverage required	70%
148097	residences served	124288
1.36%	annual increase of collection	6.87%
	year 2024 (July)	
	existing number of residences	206590
	collection coverage required	100%
	residences served	206590
	year 2027 (May)	
172640	existing number of residences	
100%	collection coverage required	
172640	residences served	

Sample calculations:

$155891x1.009^{a}=148097x1.0136^{a}$ a=11.4 years

$155891x1.009^{11.4}=172640$ $148097x1.36^{11.4}=172640$

$177554x1.02^{a}=124288x1.0687^{a}$ a=7.6 years

$177554x1.02^{7.6}=206590$ $124288x1.0687^{7.6}=205590$

2015+11.4=2026.4=May 2027 2015+7.6=2022.6=July 2023

Table 7. Strategy options after 2015. Option 2 – Reach complete collection within 12 years. Constant initial number of residences 140000

City A		City B
0.9%/year	demographic expansion	2.0%/year
90%	initial collection coverage	40%
	year 2003	
140000	initial number of residences	140000
126000	residences served	56000
	year 2015	
155891	existing number of residences	177554
95%	collection coverage required	70%
148097	residences served	124288
1.36%	annual increase of collection	6.87%
	year 2027 (end)	
173586	existing number of residences	225181
100%	collection coverage required	100%
173586	residences served	225181
1.33%	annual increase of collection	5.08%

Sample calculations:

$155891 \times 1.009^{12} = 148097 \times (b)^{12}$ b=1.0133 or 1.33%

$177554 \times 1.02^{12} = 124288 \times (c)^{12}$ c=1.0508 or 5.08%

$155891 \times 1.009^{12} = 173586$ $148097 \times 1.0133^{12} = 173586$

Upon maintaining the original effort, the first city would reach the 100% collection target in May of 2027, and the second city would reach it in July of 2023. Upon adjusting their efforts in 2015 to fit a new twelve-year deadline running until the end of 2027, the first city could relax its effort from 1.36% to 1.33% of annual collection increase, and the second city could relax it from 6.87% to 5.08%.

The option of a new twelve-year term appears to be more universal and is considered the preferred decision of the next world summit, which is likely to occur in 2012. In this context, the first two targets on the world summit time scale are now established for cities of the developing world. They are:

Target 1 for the year 2015: Reduce by one half the proportions of legal urban residences excluded from waste collection services.

Target 2 for the year 2027: Attain complete waste collection service for all legal urban residences.

This strategy is in accordance with the last and the next world summits and thus is considered definite and not negotiable. This means that after 2027 there will be no excuse for any municipal administration not having in place complete collection services. From this point in time onward, the problem of landfill diversion will be tackled by world summits and by municipal administrations. This is the general picture. Any municipality may of course advance faster, anticipate the collection targets and start diverting waste from landfills before 2027.

The Third Twelve-year Term: Moving ahead of World Summits

After having put into place complete collection practices at any point in time up to 2027, progressive municipal administrations will be looking for pragmatic landfill diversion targets. Complete collection service is the turning point on the time scale where the objective moves from taking more to taking less waste to the landfill. The target starts moving backward from 100% to 0% tipping in the long run. Having accepted the world summit time scale of twelve-year terms in the forward movement, it is reasonable to continue using it in this backward movement, too. In this context the question arises as to what should be the target for the third term on the road to zero waste that runs from 2028 to 2039, and that world summits might only address in 2022 or 20 years after Johannesburg.

The Zero Waste International Alliance has proposed a set of community principles to this effect. They "encourage communities to establish benchmarks and time lines to meet goals like diverting at least 90% of waste from landfills and incinerators within 10 to 15 years of adopting the plan" (ZWIA 2010). This reference supports the line of reasoning used in the present text, as it adopts a similar timeframe. The twelve-year period appears to be generally acceptable for municipal waste management strategies. What remains to be provided is a satisfactory explanation for the arbitrary benchmark of 90% diversion. Municipal administrators like to be convinced instead of simply being induced to believe. Consequently, the present author experimented with human behavior related to waste movement in order to determine a pragmatic and well explained target for the third interval on the time scale. The experimentation provided answers to two fundamental queries of strategy, namely

What is the best achievable waste composition after source separation?

How much coaching effort is required to achieve it?

With this type of information, a municipal administration may fix the diversion target and dimension the corresponding effort for this all important third time interval. The unsorted waste composition in developing countries was shown in Table 1. Upon intensive coaching over a period of ten years, this author was able to arrive at what is considered the best possible sorted waste composition, i.e. the composition after source separation by households in Brazil. It is shown in Table 8. The study involved a condominium complex of 48 families and a two-step separation procedure. The families were requested to separate their waste into biodegradable and inert matter, and the building employees proceeded with verification and eventual additional sorting if required.

Table 8. Composition of source separated domestic waste. Best result achievable after intensive coaching

biodegradable material appropriate for composting	47%
recyclable material appropriate for reverse logistics	20%
trash temporarily tipped at landfill or incinerated	33%

Source: Fehr et al 2010.

Inevitably, a third component appeared at the deposit area, namely unsorted waste. Here resides the secret of the theory: At this time, unsorted waste composition is not reproducible upon source separation, even after intense coaching and additional sorting by building employees. The unsorted fraction amounted to 33% in the experiments, and had to be collected and tipped. Nevertheless, 67% of all waste produced in this community was recycled. The unsorted waste composition reports in Table 1 left the impression that source separation can really produce the composition listed, which is not the case as has been shown here (Fehr et al 2010).

The argument for setting a pragmatic target for landfill diversion of domestic waste in the twelve years following the turning point is obvious from Table 8. The result in the table is experimental and therefore may be reproduced in similar or larger communities of different developing countries. The suggested target for the third interval on the world summit time scale in Brazil is landfill diversion of 67% of domestic waste. Logically, every municipal administration may choose to repeat the source separation test in order to arrive at the correct numbers for the local target, and those numbers will probably differ from one municipality or one country to another.

It remains to quantify the coaching effort. To this effect, the author has developed an experimental learning curve for lay populations (Fehr 2003). This curve provides percentages of target approximation of a given community as function of coaching effort. It also shows that 90% approximation is the best possible result. If it were desired to convince 80% of a given community to participate in source separation, the curve returns as required effort 0.5 person-years per 1000 people. For the sample community of 140000 residential units used in the previous calculations, this implies

$$(0.5 / 1000) \times 140000 = 70 \text{ person-years.}$$

This may be understood as employing 70 persons for one year, or 70/12=5.83 persons for the whole twelve-year period, or any intermediate situation. The task of those coaches would be to work in personal contact with residences to initiate and oversee the progress of source separation. This is the strategic decision. The exact procedure is a matter of tactics. The target for the third term on the world summit time scale can now be spelled out: In the course of twelve years, starting in 2028, have at least 80% of the residences practice source separation with results similar to those shown in Table 8, and provide for reverse logistics to absorb all separated biodegradable and inert items. The landfill diversion should reach approximately 67% of domestic waste.

Meeting the Challenge of Zero Waste during the Fourth and Last Interval

The previous considerations span 36 years of waste collection and diversion and moved ahead of world summits on sustainable development. Although the theory behind the targeting procedure originated from the 2002 Summit and adopted its time scale, the strategy has become independent from international directives. In fact, the relationship has been reversed. The strategy developed here is now available to be taken into account by future summit delegates.

The last interval on the proposed time scale will run from 2040 to 2051. What does it represent in terms of desired achievement? In essence, it needs to consolidate the 67% landfill diversion already reached and develop strategies to take care of the remaining 33%. This last item consists of material that source separation was not able to isolate. The reasons may be lack of attention, lack of dedication, lack of goodwill or even the fact that certain items are simply not recyclable at present. Examples are plasticized paper from food packaging, dirt from sweeping the floor, residues from the toilet and the like. At this point in time it is not obvious how to determine the best strategy. The Zero Waste International Alliance will of course recommend to move backward in the production chain and avoid producing items that later cannot be reclaimed or recycled. In this case, national political decisions are required. The municipal administrator has no authority or political clout to intervene in the circulation of merchandise, except perhaps for retail measures like prohibiting the handout of plastic shopping bags.

Thus the strategic decisions are really limited to three or four choices, such as

- Return to the learning curve and keep working on the perfection of source separation in order to move more items from the trash bin into the recycle bin.
- Consider incineration of really non-recyclable material, with all its inconveniences.
- Move up the production chain and avoid the appearance of non-recyclable material with the introduction of alternatives.
- Maintain a limited landfill to absorb residual trash items until other solutions can be found.

This type of strategic decision will depend on local conditions and administrative possibilities. No universally applicable and acceptable strategy can be indicated here. It may also be possible to develop a combined strategy that draws on all of the choices listed above. In any case, twelve years are available to satisfactorily define what is meant by zero waste in a particular city and to establish the strategy to reach the target.

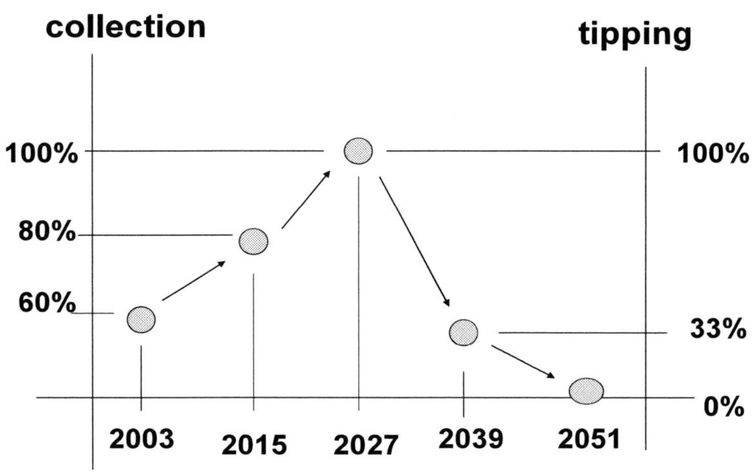

Figure 1. The target plot for the basic initial situation.

The Strategy in Form of the Target Plot

As a result of this research, the target plot shown in Figure 1 is put forth to represent the basic strategy of approaching a zero waste situation in the municipal context on a time scale of 48 years extrapolated from world summit directives. The four twelve-year terms discussed in the text are shown in the figure with their respective targets. As the turning point is reached, the left ordinate is replaced by the right ordinate to mark the progress.

The Most Appealing Tactical Procedures

Having established the strategy, it becomes a matter of tactics to work toward the targets. Just as was the case with strategy, the tactics also vary from country to country. In the case of the strategy, it was mentioned that the first two targets relating to complete collection service are not likely to apply to developed countries, where they have been met in the past. As for tactics, the ideas of successful procedures differ again. The construction of mathematical models to determine the location of drop-off areas and to assess the willingness to cooperate, are of limited value in developing countries.

From the author's experience in Brazil, three basic tactical approaches are suggested to municipal administrations in developing countries. They refer to the three main parts of domestic waste management, namely source separation, biodegradable waste handling and inert waste handling, and may be spelled out as follows:

1. Use the learning curve to provide the right amount of residential coaching.
2. Create effective reverse logistics for biodegradable matter.
3. Subsidize certain inert trash items for limited time periods.

The learning curve, as mentioned before, indicates a required coaching intensity of at least 70 person-years to put source separation into practice in 80% of 140000 residences. What would be expected of each coach? If 70 persons are assigned to the task for one year, the arithmetic very simply states 140000/70/365=5.5 residences per day per person. This means that each coach has to visit 5.5 residences per day during that year to get the program under way. The coaches may be employees of the public administration, but even better, may be recruited from the student universe of schools in the city. Apart from this individual coaching, collective programs are recommended to convey the targets and obtain commitments. As examples, associations of retailers and condominium complexes in the city may be involved in the effort without cost to the public budget. The idea here is to create precedents in a few condominium buildings and shops that can later be imitated. Legal support for this undertaking is important. A municipal bylaw that enforces source separation would facilitate the coaching.

The reverse logistics for biodegradable material is inexistent at present, but represents the most important component of the landfill diversion effort because approximately 70% of all domestic waste falls into this category. With the funds liberated by reduced tipping costs, incentives can be created for small or large scale composting operations. As source separation progresses in the city, pilot composting units may be installed and operated by private enterprises. The bottleneck occurs in the collection. Appropriate containers will have to be found. In the ideal case, and this derives from the authors own experience, the waste movement proceeds in closed containers. The residences fill them and hand them over to the collection crew who leaves empty containers in exchange. At the composting site, the material is integrated into the windrows and the containers are washed and returned. The city administration has to see to it that a market exists for the compost, most likely by subsidizing the price with funds from reduced landfill costs. In any case, as has been mentioned before, tipping of biodegradable waste has no future, regardless of the cost.

The inert part of domestic waste is the easiest to manage, because informal reverse logistics is already well developed to handle it. The secret resides in stimulating this reverse logistics with simple but effective actions. At present, Brazil as a country prides itself of being one of the world champions in recycling aluminum cans. How did this happen? The administration nothing contributed. The market took care of the circuit. Used aluminum cans in 2010 are commercialized at the wholesale level for USD1294 per ton. Consequently, any can thrown away at any place in town will have been picked up by somebody within minutes. Hardly any can is left for the scavengers to collect. The question is: Why can this phenomenon not be replicated for other items? The answer is: The prices are not conducive. So the tactics become evident. Cardboard e.g. is traded at wholesale level for USD265 per ton. Consequently, nobody bothers picking up and selling cardboard. Whatever scavengers do not take away ends up in the landfill, and there is lots of cardboard sitting on sidewalks every day. The same is true for plastic bottles, steel cans and glass, which sell at USD771, USD323 and USD118 per ton respectively. The cost to the city budget for collecting and tipping those items may run in the order of USD165 per ton, depending on the type of contract with third parties. It is easy to see that there are ample opportunities for stimulating recycling operations through the transfer of funds between accounts.

Conclusion

The time scale established by the 2002 World Summit on Sustainable Development can be extrapolated to define progress targets of waste management in municipalities of the developing world for the next five decades.

The effort required to reach waste management targets depends on the initial collection coverage and the demographic expansion.

This effort has been quantified in terms of annual progress targets.

Directives to be elaborated by future world summits are anticipated and proposed as guides for municipal administrations.

A generally applicable targeting plot is presented for reaching a zero waste situation from any initial condition within five decades.

Key tactical parameters of procedure are discussed. They refer to source separation and reverse logistics for both biodegradable and inert waste components.

The administrative vacuum referring to the strategy of target plotting has been removed, and the 48-year timeframe for approaching zero waste is realistic because its scope is perfectly within the reach of administrative resources.

This is the first time that a universal and realistic targeting strategy with a pragmatic timeframe applicable in the developing world is being presented.

References

Aini, M.S., Fakhru'l-Razi, A., Lau, S.M., Hashim, A.H. 2002 *Practices, attitudes and motives for domestic waste recycling* International Journal of Sustainable Development and World Ecology, Parthenon Publishing Group UK, vol.9 no. 3 pp. 232-238.

Ando, A.W., Gosselin, A.Y., 2005 *Recycling in multifamily dwellings: Does convenience matter?* Economic Inquiry, Wiley-Blackwell International, vol. 43 no. 2 pp. 426-438.

Atasoylu, G., Ergin, F., Evci, E.D., Tikir, D., Kaya, E., Beser, E. 2007 *The household garbage in the western coast region of Turkey and its relationship with the socio-economic characteristics*, Journal of Environmental Biology, Triveni IN, vol. 28 no. 2 pp. 225-229.

Basili, M., DiMatteo, M, Ferrini, S. 2006 *Analyzing demand for environmental quality: A willingness to pay/accept study in the province of Siena, Italy*, Waste Management, Elsevier International, vol. 26 no. 3 pp. 209-219, ISSN 0956 053X.

Beatty, T.K.M., Berck, P., Shimshack, J.P. 2007 *Curbside recycling in the presence of alternatives*, Economic Inquiry, Wiley-Blackwell International, vol. 45 no. 4 pp. 739-755.

Benítez, S.O., Vega, C.A., Barreto, M.E. 2003 *Characterization and quantification of household solid waste in a Mexican city*, Resources, Conservation and Recycling, Elsevier International, vol. 39 no. 3 pp. 211-222 ISSN 0921 3449. http://dx.doi.org/10.1016/S0921-3449(03)00028-4.

Bohm, R., Folz, D., Kinnaman, T.C., Podolsky, M. 2010 *The costs of municipal waste and recycling programs*, Resources, Conservation and Recycling, Elsevier International, vol. 54 no. 11 pp. 864-871 ISSN 0921 3449.

Buenrosto, O., Bocco, G. 2003 *Solid waste management in municipalities in Mexico: Goals and perspectives*, Resources, Conservation and Recycling, Elsevier International, vol. 39 no. 3 pp. 251-263 ISSN 0921 3449.

Calçado, M.R. 1998, *Household waste: from proposal to tests of a proactive management model* (in Portuguese), Master of Science Dissertation 120 pp., Federal University at Uberlândia, Brazil, library code 628.54 C144r TES/FU, http://www.bibliotecas.ufu.br/catalogo > termos de pesquisa: Calçado, Marilda.

Carmo, M.S.D., Oliveira, J.A.P. 2010 *The semantics of garbage and the organization of recyclers: implementation challenges for establishing recycling cooperatives in Rio de Janeiro*, Resources, Conservation and Recycling, Elsevier International, vol. 54 no. 12 pp. 1261-1268 ISSN 0921 3449.

Castro, M.S.M.V. 1998, *A comparative analysis of the household waste management scheme used in Uberlândia* (in Portuguese), Master of Science Dissertation 122 pp., Federal University at Uberlândia, Brazil, library code 628.4.032 C355a TES/FU http://www.bibliotecas.ufu.br/catalogo > termos de pesquisa: Castro, Mirlei.

Chatterjee, R. 2009 *Municipal solid waste management in Kohima City, India*, Waste Management, Elsevier International, vol. 29 no. 11 pp. 2909-2910 ISSN 0956 053X

Chen, C.C. 2010 *A performance evaluation of MSW management in Taiwan*, Resources, Conservation and Recycling, Elsevier International, vol. 54 no. 12 pp. 1353-1361 ISSN 0921 3449.

Dangi, M.B., Pretz, C.R., Urynowicz, M.A., Gerow, K.G., Reddy, G.M. 2011, *Municipal solid waste generation in Kathmandu, Nepal*, Journal of Environmental Management, Elsevier International, vol. 92 no. 1 pp. 240-249 ISSN 0301 4797. http://dx.doi.org/10.1016/j.jenvman.2010.09.005

Dijkgraaf, E., Gradus, R.H.J.M. 2004 *Cost savings in unit-based pricing of household waste: The case of the Netherlands*, Resource and Energy Economics, Elsevier International, vol. 26 no. 4 pp. 353-371.

Dresner, S., Ekins, P. 2010 *Charging for domestic waste in England: Combining environmental and equity considerations*, Resources, Conservation and Recycling, Elsevier International, vol. 54 no. 12 pp. 1100-1108 ISSN 0921 3449.

EFN 2010 Ecological Footprint Network http://www.footprintnetwork.org (2010 12 26).

Eriksson, O., Baky, A. 2010 *Identification and testing of potential key parameters in system analysis of MSW management*, Resources, Conservation and Recycling, Elsevier International, vol. 54 no. 12 pp. 1095-1099 ISSN 0921 3449.

Ewing, G. 2001 *Altruistic, egoistic and normative effects on curbside recycling*, Environment and Behavior, SAGE Journals, Vienna AT, vol. 33 no. 6 pp. 733-764.

Falkner, R., Stephan, H., Vogler, J. 2010 *International Climate Policy after Copenhagen: Towards a building block approach*, Global Policy, Wiley-Blackwell International, vol. 1 no. 3 http://dx.doi.org/10.1111/j.1758-5899.2010.0004J.x

Fehr, M. 2002 *The prospect of municipal waste landfill diversion depends on geographical location* The Environmentalist, Springer International vol. 22 no. 4 pp. 319-324 ISSN 0251 1088 eISSN1573-2991 http://dx.doi.org/10.1023/A:1020710829477.

Fehr, M. 2003 *Environmental management by the learning curve* Waste Management, Elsevier International vol. 23 no. 5 pp. 397-402 ISSN 0956 053X http://dx.doi.org/10.1016/S0956-053X(03)00063-1.

Fehr, M., Castro, M.S.M.V. 1999, *Garbage analysis infers management model* (in Portuguese) Saneamento Ambiental, Signus Editora, São Paulo BR vol. 10 no. 55 pp. 38-41 ISSN 0103 7056 http://www.signuseditora.com.br

Fehr, M., Calçado, M.R. 2001, *Divided collection model for household waste achieves 80% landfill diversion*, Journal of Solid Waste Technology and Management, Widener University, Chester US, vol. 27 no. 1 pp. 22-26 ISSN 1088 1697 http://www2.widener.edu/~sxw0004/abstracts.html

Fehr, M., Castro, M.S.M.V., Calçado, M.R. 2010 *Condominium waste management by private initiative: a report of a ten-year project in Brazil*, Waste Management & Research, The official journal of ISWA, SAGE Journals, Vienna AT, vol. 28 no. 4 pp. 309-314, ISSN 0734 242X eISSN 1096 3669 http://dx.doi.org/10.1177/0734242X09344337.

Ferrara, I., 2003 *Differential provision of solid waste collection services in the presence of heterogeneous households*, Environmental & Resource Economics, Springer International, vol. 26 no. 2 pp. 211-226.

Ferrara, I., Missios, P. 2005 *Recycling and waste diversion effectiveness: Evidence from Canada*, Environmental & Resource Economics, Springer International, vol. 30 no. 2 pp. 221-238.

Gallardo, A., Bovea, M.D., Colomer, J., Prades, M., Carlos, M. 2010 *Comparison of different collection systems for sorted household waste in Spain*, Waste Management, Elsevier International, vol. 30 no. 12 pp.2430-2439, ISSN 0956 053X.

García, A.J., Esteban, M.B., Marquez, M.C., Ramos, P. 2005 *Biodegradable municipal solid waste: characterization and potential use as animal feedstuffs*, Waste Management, Elsevier International, vol. 25 no. 8 pp. 780-787, ISSN 0956 053X.

Guimarães, B., Marques, R.C., Simões, P. 2010 *Does performance evaluation help public managers? A balanced score card approach in urban waste services*, Journal of Environmental Management, Elsevier International, vol. 91 no. 12 pp. 2632-2638, ISSN 0301 4797.

Hong, J., Li, X., Zhaojie, C. 2010 *Lifecycle assessment of four MSW management scenarios in China*, Waste Management, Elsevier International, vol. 30 no. 11 pp. 2362-2369 ISSN 0956 053X.

Jenkins, R.R., Martínez, S.A., Palmer, K., Podolsky, M.J. 2003 *The determinants of household recycling: a material-specific analysis of recycling program features and unit pricing*, Journal of Environmental Economics and Management, Elsevier International, vol. 45 no. 2 pp. 294-318.

Kipperberg, G. 2007 *A comparison of household recycling behavior in Norway and the United States*, Environmental and Resource Economics, Springer International, vol. 36 no. 2 pp. 215-235.

Lavee, D., Khatib, M. 2010 *Benchmarking in MSW recycling*, Waste Management, Elsevier International, vol. 30 no. 11 pp 2204-2208 ISSN 0956 053X.

Morris, J. 2005 *Comparative LCAs for curbside recycling versus landfilling or incineration with energy recovery*, International Journal of Life Cycle Assessment, Ecomed Publishers, vol. 10 no. 4 pp. 273-284.

Nabegu, A.B. 2010 *An analysis of municipal solid waste management in Kano Metropolis, Nigeria*, Journal of Human Ecology, Kamla-Raj Enterprises, Delhi IN, vol. 31 no. 2 pp. 111-119, ISSN 0970 9274.

Qu, X.Y., Li, Z.S., Xie, X.Y., Sui, Y.M., Yang, L., Chen, Y. 2009 *Survey of composition and generation rate of household waste in Beijing*, Waste Management, Elsevier International, vol. 29 no. 10 pp. 2618-2624 ISSN 0956 053X.

Sembiring, E., Nitivattananon, V. 2010 *Sustainable solid waste management toward an inclusive society: Integration of the informal sector*, Resources, Conservation and Recycling, Elsevier International, vol.54 no. 11 pp. 802-809 ISSN 0921 3449.

Shinkuma, T. 2003 *On the second-best policy of household's waste recycling*, Environmental and Resource Economics, Springer International, vol. 24 no. 1 pp. 77-95.

Tralhão, L., Coutinho-Rodrigues, J., Alçada-Almeida, L. 2010 *A multi-objective modeling approach to locate multi-compartment containers for urban solid waste*, Waste Management, Elsevier International, vol. 30 no. 12 pp. 2418-2429 ISSN 0956 053X.

UNEP 2010 United Nations Environmental Program http://www.unep.or.jp/ietc/SPC/Publications/asp (2010 12 17).

United Nations 2002 Johannesburg Summit http://www.un.org/events/wssd (2010 11 27).

United Nations 2000 Millennium Goals http://www.un.org/millenniumgoals (2010 11 23).

Yan, Y. 2010 *Domestic waste recycling, collective action and economic incentives: The case of Hong Kong*, Waste Management, Elsevier International, vol. 30 no. 12 pp. 2440-2447 ISSN 0956 053X.

Yang, H.L., Innes, R., 2007 *Economic incentives and residential waste management in Taiwan: An empirical investigation*, Environmental & Resource Economics, Springer International, vol. 37 no. 3 pp. 489-519.

ZWIA 2010 Zero Waste International Alliance http://www.zwia.org (2010 12 13).

In: Municipal Solid Waste
Editors: G. Konno and C. Machado

ISBN 978-1-62100-257-4

Chapter 2

SYNCHRONIZATION OF SOLID WASTE MANAGEMENT IN URBAN CENTERS WITH INTEGRATED NUTRIENT MANAGEMENT IN INDIAN AGRICULTURE

J. K. Saha[1,*], N. Panwar[2] and M. C. Manna[1]

[1]Indian Institute of Soil Science, Nabibagh, Bhopal –M.P., India
[2]Central Arid Zone Research Institute, Jodhpur, Rajasthan, India

ABSTRACT

Organic matter status of cultivated lands in most of the agroecological regions of India is low due to rapid mineralization loss of C input into the soil from different sources. This, in conjunction with low fertilizer input, has resulted in continuous decline in agricultural productivity. Long-term fertilizer experiments as well as manurial trials in various parts of the country have conclusively shown that integrated nutrient management (INM) involving various types of organic manure can sustain or augment agricultural productivity. Rural compost prepared mainly from cattle shed litters and agricultural wastes remained major source organic matter in the INM in Indian agri-horticulture. However, availability of rural compost is very low (on average, little over 2 t/ha) and is declining continuously due to decreased dependency on animals in crop husbandry.

On the other hand, India generates about 70 million tonne (MT) of municipal solid wastes annually, which has the potential of producing about 9 MT compost (MSWC). This has the huge potential of supplying important plant nutrients N (55 thousand tonne / year), P (14 thousand tonne / year), K (40 thousand tonne / year) and Zn (2.3 thousand tonne / year) to cultivated land besides providing organic matter for improvement of physical and biological environment. A survey in 29 cities has indicated that quality of MSWC in respect of fertilizing potential and heavy metals contents varied widely with the processes adopted for manufacturing. In general, fertilizing potential of the MSWC in majority of the cities is very low, resulting in their poor marketing. Such a situation has ultimately caused poor volume (presently about 9%) of recycling of solid wastes into compost. This review intends to deal with three issues; namely, demand of organic matter for INM in agriculture, recycling

* Corresponding author. E-mail: jk_saha12000@yahoo.com; Telephone: +91-755-2730970; Fax: +91-755-2733310.

of MSW towards compost manufacturing, and protection of soil resources from heavy metal pollution. A new system of classification for MSWC into different marketable classes (A, B, C, and D) and restricted use classes (RU-1, RU-2, and RU-3) has been proposed on the basis of their fertilizing potential and heavy metal pollution potential. India has diversified cropping system which includes foodgrain crops, non-food crops, and horticultural crops covering about 71%, 23% and 6% of total gross cropped area. Potential use for each of the classes of MSWC in INM of different agri-horticulture systems has been discussed keeping in view of nutrient requirement as well as probability of contamination of environment. This classification system is expected to maximize recycling of MSW, reduce the burden of land-fill sites, reduce the gap between demand and supply of organic matter for INM, improve soil physical environment and reduce environmental pollution by mitigating emission of methane (a green house gas) from landfill sites. Ways of improving present system of solid waste management for producing good quality compost have also been suggested.

INTRODUCTION

Like many other developing countries, India has achieved commendable success in achieving self-sufficiency in production and productivity of major crops. The untiring efforts of scientists, farmers, policy makers and input agencies helped in changing scenario of the country from begging bowl image during 1950s to that of reasonable surplus and exports during 1970s. Introduction of inorganic fertilizer nutrients had significant contribution to such rapid progress in the agricultural sector. However, recent energy crisis and consequent price hike of fertilizer due to withdrawal of subsidy on fertilizer coupled with the low purchasing power of the farming community have again reviewed interest in organic recycling in the India as well as other developing countries. Since past 30 years or so, research on nutrient management of crops had been gradually oriented towards integrated use of organic resources along with fertilizers; where, well composted manure prepared from farm wastes remained at the focus and has been adopted well in the farmers' practice. However, due to low availability of such compost vis-à-vis its requirement, utilization of organic matter from other sources is being emphasized upon. Several experiments (both short-term and long-term in duration) have been conducted in developed countries on the utilization of urban waste compost in agriculture and had generated information on its potential as well as threat on environment (Hargreaves et al., 2008). However, comprehensive information in respect of recycling municipal solid wastes (MSW) for its beneficial use in agriculture is very scanty in India. This article attempts to understand the obstacle in adoption of MSW compost in the integrated nutrient management in Indian agriculture as well as to suggest doing away with it for better utilization.

NUTRIENT REQUIREMENT OF INDIAN AGRICULTURE

India has a total geographical area of 328.7 m ha. The gross cultivated area forms about 58 per cent of the total reporting area. In some states such as Punjab and Haryana the net sown area is about 75 and 83 percent of the reporting area and in West Bengal the value is more than 60%. Due to its high variability in climate, water availability, soil and land morphology, India has quite vast range of crops and cropping systems. These ranges from annual food grain crops; cereals, pulses, vegetables, oilseed crops; non-food crops like cotton, jute, tobacco; plantation crops; fruit crops; condiments and spices (Table 1). Nature, amount and mode of nutrient requirement of these crops vary widely.

Contribution of fertilizers to total grain production was about 1% in 1950, 33% in 1980 and 53% in 2000. The growth in fertilizer consumption has come down during 1990's and there is stagnation like situation for the last 4-5 years. At the present level of food production, the nutrients requirement is estimated as about 32 million tonnes and supply through fertiliser is about 26 million tones. Thus a deficit of nutrient remains around 6 million tonnes. This deficit is to be met from renewable source of nutrients i.e., organic and biological sources of nutrients. Nitrogen can be gained from both sources; whereas, P and K would be mainly obtained from recycling of organic matter and crop residues.

The stagnation in fertilizers consumption in recent years due to significant rise in their price and higher negative nutrient balance are posing a threat to soil health and sustainable agriculture. Moreover, farmers apply chemically pure form of fertilizers containing mostly major nutrients N, P and K. This resulted continuous mining out of essential micronutrients as well as sulphur from soil without their supplementation through external sources. Analysis of 150 thousand soils samples from different regions of the country indicated widespread deficiency of Zn (47%), B (20%), Mo (18%) and Fe (12%). Emerging deficiencies of micronutrients are other major factors of deteriorating soil quality and declining in factor productivity of nutrients. Use of micronutrient containing fertilizers is very limited due to either lack of awareness or non-availability.

INTEGRATED NUTRIENT MANAGEMENT FOR SUSTAINABILITY OF CROP PRODUCTION

Integrated nutrient management (INM) is a concept that aims at managing and supplying nutrient to plants through various sources such as nutrient reserves in soil as well as inorganic fertilizer and organic sources to an optimum level for sustaining desired crop productivity. This optimizes the benefit from all possible sources of plant nutrients in an integrated manner. Intensive agriculture with low and imbalanced supply of plant nutrient has resulted deterioration of native soil fertility posing a serious threat to long-term sustainability of different production systems (Swarup, 1998). The interactive advantage of combined use of all possible sources of nutrients and their scientific management in INM has proved superior to use of its each component for optimum growth, yield and quality of different crops and cropping systems as well as maintaining or improving soil quality parameters.

Role of Organic Manure in INM

Rural waste compost has been incorporated into INM for intensive cropping sequences during the last 30-35 years. The potential nutrient availability from major organic sources is estimated at 10.5 to 16.2 m tonnes, out of which only 3.9 to 5.7 m tones is available for agricultural use.

Table 1. Area under different crops in India during the year 2009 (FAI, 2010)

Sl. No.	Crops	Area (thousand ha)
1.	Food grain crops	122832.4
2.	Oilseed crops	27557.8
3.	sugarcane	4415.4
4.	Cotton	9406.7
5.	Jute	785.6
6.	Fruits	6477.8
7.	Vegetables	8214.6
8.	Tea	579.3
9.	Coffee*	278.6
10.	Rubber	687.0
11.	Spices & condiments**	7272.0

* for the year 1991-'92; ** for the year 2007-'08.

Field trials were conducted on 66.7 ha of farmers' fields in rainfed area of 34 villages in 7 states on integrated nutrient management for predominant oilseed based cropping system prevailed in the respective area (Saha, 2003). Results showed that inclusion of organic sources of nutrients in the INM enhanced yield of oilseed crops by 6 to 31% over recommended doses of fertilizer nutrients (Table 2). Integrated nutrient management treatments involving different sources of organic manure maximized yield as well as income in different oilseed based cropping system (Table 3). For example, application of farm-yard manure (FYM) in soybean-chickpea system and in fallow-sunflower system; lime along with FYM in groundnut + pigeonpea intercropping system; green manuring in safflower, castor and mustard were found beneficial in increasing yield as well as income over farmers practice as well as recommended doses of fertilizers (RDF).

Table 2. Yield advantage of oilseed crops due to INM treatments over Farmers practice

Oilseed crops	Yield with farmers practice (q/ha)	Yield with 100% RDF (q/ha)	Increase over farmers' practice (%)	Yield with best INM (q/ha)	Increase in INM over RDF (%)
Soybean	8.47	10.52	24.2	11.71	11
Safflower	11.59	12.54	8.2	14.43	15
Sunflower					
At Latur	5.07	5.55	9.5	7.17	29
At Raichur	7.15	8.82	23.3	10.06	14
Groundnut	5.94	8.74	47.1	10.71	23
Castor	4.52	5.69	25.9	6.02	6
Mustard	16.51	19.14	15.9	20.88	9
Raya	6.75	7.55	11.8	9.90	31

Table 3. Economics of the cropping system

Cropping system	Additional return* due to RDF (US $/ ha)	Additional return* due to best INM (US $/ ha)	Additional return due to best INM over RDF (US $/ ha)
Soybean - Chickpea	85.1	136.0	50.9
Greengram - Safflower	13.3	68.1	54.8
Fallow - Sunflower			
At Latur	8.9	53.2	44.3
At Raichur	44.3	72.6	28.3
Groundnut + pigeonpea	254.3	370.4	116.1
Castor monocropping	21.2	32.2	11.0
Fallow-Mustard	56.5	95.2	38.7
Maize-Raya	17.0	57.6	40.7

* With respect to farmers practice.

Data from long-term experiments have revealed that additional yields of different crops could be realized over and above soil test based optimum rates of N, P and K if only 10-15 t /ha of FYM was annually supplemented along with NPK doses (Table 4). Average yield with 100% recommended doses of N, P & K plus FYM was even more than the average yield obtained with 150% NPK. Besides increasing yield of crops, continuous application of FYM has been found to improve soil quality parameters in different agro-ecoregions.

Besides increasing crop yields, organic manure application resulted enhancement of soil fertility. In Ranchi, continuous application of FYM for 28 years improved soil pH, and resulted higher available soil moisture content and lower bulk density of soil as compared to the treatments supplying nutrients only through inorganic fertilizers (Table 5).

While continuous application of fertilizers results depletion due mining out by crops, organic manure addition sustains or even builds up micronutrient reserve in soil. Application of FYM for 27 years in maize-wheat cropping system on alfisol increased available Zn and Cu content in soils (Lal and Mathur, 1989). Improvement in soil physical and chemical properties of soils as well as yields of sugarcane were noticed with integrated nutrient management involving FYM in 10 years study in Udic Haplustalf at Mandya (Rabindra and Gowda, 1986). In a 45 years study on a vertisol (Pune) seed cotton and sorghum yields, SOM status, available N, P, K and maximum water holding capacity were significantly higher in FYM (@ 6.2 t /ha) applied plots compared to unmanured plots (Table 6).

Table 4. Average grain yield of crops (t ha^{-1}) over the years in long-term experiments on yield stability and productivity (Swarup, 1998)

Soil type	Control	NPK 100%	NPK 150%	NPK 100% + FYM
Inceptisol	1.13	3.13	3.50	3.64
Vertisol	1.05	3.25	3.35	3.50
Mollisol	2.50	4.65	4.80	5.40
Alfisol	0.55	2.73	3.12	3.37

Table 5. Changes in soil properties of an alfisol after 14 and 28 years due fertilizer and manure treatments (Sarkar et al., 1989)

Treatment	Bulk density (g /cm^3)		Water-holding capacity (%)	
	1970	1984	1970	1984
Check	1.44	1.54	31.8	33.3
N	1.46	1.54	38.3	39.3
NP	1.40	1.51	37.4	36.2
NPK	1.36	1.46	33.9	40.5
NPK + lime	1.48	1.50	31.6	42.4
FYM	1.43	1.32	33.5	47.2
Initial (1956)	*1.45*		*31.5*	

Table 6. Physico-chemical properties of a vertisol and mean yield of cotton and sorghum grown in rotation after 45 years under rainfed condition (Khiani and More, 1984)

	No manure	FYM @ 6.2 t /ha	$LSD_{0.05}$
Seed cotton (t /ha)	0.15	0.31	0.03
Sorghum grain (t /ha)	0.84	1.08	0.22
Organic C (%)	0.56	1.14	0.04
Total N (%)	0.05	0.06	0.01
Available N (ppm)	75.0	92.3	2.26
Available P (ppm)	11.1	14.5	0.56
Available K (ppm)	290	333	47
Water-holding capacity (%)	57.3	64.3	1.3
Moisture retention at 0.33 bar (%)	42.8	44.8	2.6

About 200 trials conducted on farmers' fields on rice-wheat system under All India Coordinated Agronomic Research Project have shown that combined use of 12 t FYM /ha and 60 kg N /ha produced rice grain yield equivalent to that obtained with 120 kg N /ha (Kulkarni et al., 1978).

Solid Waste Generation in Indian Cities

About 7935 towns of India, harboring 31.2% of country's population generate more than 70 million tones of municipal solid wastes (MSW). A CPCB sponsored study conducted by NEERI on 59 selected cities of India revealed that waste generation rate varied from 170 to 620 g per capita per day with an average value of 400 g per capita per day (Kumar et al., 2009). Generally, more populous cities generates higher amount of wastes per capita per day. Cities with more than one lakh population contributed 72.5% of the waste generated in the country as compared to other 3955 urban centers that produce only 17.5% of that total waste (MOUD 2005).

Table 7. Composition of un-segregated MSW collected from different cities (Average of six cities)

Fraction	Content (%)	
	Range	Average
Plastics	2.1-5.4	4.1
Glassware	3.1-6.5	4.5
Rag	0.3-3.9	2.8
Metals	0.5-4.5	1.7
Stone/soil	24.9-39.7	30.7
Compostable materials	35.2-47.1	39.8
Rubber	2.3-4	3.3
Others (ceramic, earthen pot)	11.9-18.5	13.2

Solid wastes in the city area are mostly generated from domestic activities, market places, restaurants, garden, parks, street sweeping, construction & demolition activities as well as from industries operating within municipal limits. These contain recyclables (paper, plastic, glass, metals etc.), compostable organic matter (food waste, fruit and vegetable peels, woods, animal dead bodies), toxic substances (paints, used batteries, electrical and electronic spares, pesticides, medicines etc.) and soiled wastes (hospital wastes containing blood stained cotton, sanitary napkins, disposable syringes). Solid wastes collected by urban local bodies are seldom segregated at the point of origin. In most of the cities, mixed wastes are collected in community bins and at some places, a fraction of recyclables (paper, cardboard and plastics) are segregated out from these bins by rag pickers (Kumar et al., 2009). A study was carried out to investigate into the composition of non-segregated MSW generated from six cities of India. Results (Table 7) showed that composition of non-segregated MSW varied widely and contained relatively low (<50%) biodegradable material.

Characterization studies carried out much earlier by NEERI (1995) also indicate that MSW contained variable organic fraction (30-40%), calorific value of refuse ranges between 800-1000 kcal/kg and C/N ratio ranges between 20 and 30.

Existing Solid Waste Management Scenario

In most of the cities, the Municipal solid waste (MSW) management system comprises mainly three activities; waste collection, transportation and disposal. Disposal of these wastes has always been perceived as great challenge by the municipal authorities/local bodies as it involves land, labour and capital besides environmental issues. More than 90% of MSW is directly disposed of on the land in open dump without taking any precautions which pose environmental hazards. In order to enhance the carrying capacity of old landfill area /dumping ground, MSW is burnt occasionally causing severe air pollution. It is estimated that unscientific disposal of MSW releases about 604.5 thousand tonnes of methane which constitute about 0.73% of total green house gas (GHG) emission from the country (MOEF, 2010).

Composting of Solid Waste: A Viable Technology for Beneficial Recycling

Composting is the controlled and accelerated decomposition of organic matter through biological processes, resulting in nutrient-rich humus. Quality of compost is generally evaluated in terms of its ability to supply plant nutrients as well as to improve soil physical environment in favor of congenial rhizosphere. This depends mainly on the amount and composition of organic matter used as input material during the process. Composting has a long tradition in India (Howard, 1943) and is particularly widespread in the rural areas, where it uses mainly crop residues and cattle shed litter. Although Ministry of Environment and Forest, Govt. of India put forward Municipal Solid Wastes (Management and Handling) rules, 2000 and indicated waste processing options including segregation and composting of biodegradable wastes, these were complied in very few municipalities (CPCB, 2005). Presently, a very small part (about 10%) of the MSW generated is used for compost production by various public or private enterprises. Even, methodology adopted by different manufacturers for composting varies widely in respect of collection and processing of wastes prior composting, design of composting area, use of microbial culture, extent and method of aeration provided to the composting material, period of composting, handling and processing of matured composts etc. (Saha et al. 2008). Considering an average collection efficiency of about 70%, country has the potential of producing about 10 million tons of compost each year if such wastes are properly used for composting.

Survey of 36 MSW compost manufacturing plants from 29 cities of 13 states of the country indicated (Saha et al., 2008) that windrow method of composting is followed by about 72% of the manufacturers, where mostly non-segregated or partially segregated wastes are allowed to decompose for several months with or without turning. Composting was performed in concrete pits in 4 sites; viz, Kullu, Manali (passively aerated), Vijaywada and Namakkal (vermicomposting). For proper and speedy decomposition of biodegradable segment of the waste, addition/spraying of inoculants of microbes are done in majority of the composting plants. In 23% manufacturing sites, vermicomposting is being adopted for producing good quality composts in shorter time.

Pre-processing generally involves segregation of biodegradable organic fractions from the rest of the wastes, mainly composed of paper, plastics, glass, metals etc. While biodegradable organic fractions are taken to composting yards, remaining fractions are either dumped in landfill site or are segregated further to remove recyclable materials like papers, plastics and glass etc., as followed by Suryapet municipality of A.P. Pre-processing of wastes is not followed at all in majority (about 45%) of the sites where non-segregated wastes are dumped in heaps and left for several months with or without turning for decomposition of biodegradable wastes (termed as 'non-segregated'). In some of the composting plants (26%), big size non-biodegradable wastes, like papers, plastics, rubber, metals etc. are manually removed prior to composting (termed as 'partially segregated'). In rest of the cases, biodegradable wastes are segregated at generator level and this is done in two ways. In few cities (Namakkal, Suryapet and Vijaywada), individual households delivers segregated biodegradable wastes separately during 'door-to-door' collection by municipal organization which are composted in pits or windrows using earthworms. In several other cities, sources generating mainly biodegradable wastes, like hotels & restaurant, vegetable market, slaughter house etc. are selected for feedstock collection for composting plants. Wastes collected through these two methods for compost preparation will be subsequently termed as 'segregated wastes'. In many of the composting yards, microbial cultures are also added to the wastes in order to hasten the process of composting. Post-processing methods mainly involve air-drying of the composts followed by their sieving, either mechanically or manually to remove bigger size inerts. Mechanical composting plants are installed by big enterprises in many cities like Kolkata, Nashik, Bhopal, Delhi, Shimla and Solan. However, in many cases, machinery and other equipment are larger in their handling capacity than required for the quantity of wastes generated in the city.

Manurial Potential of MSW Composts Produced in Indian Cities

An insight of the results of 36 samples collected from 29 cities indicates that the physical, chemical and nutritional characteristics in the collected samples varied widely among the compost samples collected (Saha et al., 2010). The values varied from 3.6 to 45.4 % for moisture, 0.52 to 1.15 g/cc for bulk density, 6.64 to 9.63 for pH (1:5), 0.52 to 8.41 dS /m for EC, 5.2 to 22.6 % for total organic carbon, 0.26 to 1.71 % for total nitrogen, 0.08 to 0.73 % for total phosphorous and 0.12 to 1.31 % for total potassium. Minimum expected values or range of these parameters for composts have been recommended as 6.5 – 7.5 for pH (1:5), < 4 dS /m EC, 16 - 20 % for total organic carbon, 0.8 % total nitrogen, 0.22 % total phosphorous and 0.83 % total potassium (FAI, 2007). Considering these limit values 35 % samples are having higher pH (i.e. >7.5) and four samples are having higher EC (i.e. >4 dS /m). Almost all the samples (85%) are below the expected values for total organic carbon. Out of total MSW compost samples 38, 20 and 9 % samples contained minimum expected content of total N, P and K, respectively.

Segregation of wastes prior to composting had significant effect in several important chemical parameters of the finished product. Moisture content, EC, Cl^-, oxidizable and total organic C, total N, total P contents in the MSW composts produced from segregated solid wastes were respectively about 90, 59, 65, 54, 39, 64 and 75% higher as compared to these values in the composts produced from non-segregated solid wastes. Though, partial segregation of wastes before composting did also improve the quality parameters; but the magnitudes of increases were not statistically significant.

Scope of MSW Composts in Integrated Nutrient Management

Several researches, both in India and abroad indicated that MSW compost had certain beneficial effects on soil quality and had the potential of increasing crop yields. A considerable volume of research has been conducted in developed countries through long-term experiments on utilization of MSW compost in agriculture and their application benefits in crop production and soil fertility improvement has been elaborately reviewed by Hargreaves et al. (2008). In majority of the long-term studies with high application rate, compost improved physical properties, fertility status and biological health of soil through concurrent enhancement of organic matter content. However, such information through long-term studies are practically absent in India. Further, most of the studies were conducted using smaller dose of application of poor quality of compost prepared from non-segregated wastes. Field experiments at Kolkata (India) indicated that MSWC application for two years improved yield of rice and was safe to be used as an alternative organic supplement (Bhattacharyya et al., 2003). Application of MSW compost along with urea increased rice grain yield by 49%. Due to low content of organic matter and plant nutrients, MSW compost was not as effective as cattle dung manure in increasing rice grain yield. As mechanical compost plant at Kolkata use non-segregated wastes as input material, compost produced were of inferior quality. Similarly at Coimbatore (India), application of MSW compost @ 4.7 t/ha saved 25% of the recommended dose of N in rice crop (Kavitha and Subramanian, 2007). These studies clearly indicate that even inferior quality of MSW compost can beneficially be used in INM along with fertilizer nutrients and can save costly fertilizer nutrients without any compromise on yield of crops.

Results, thus, indicate that even inferior quality of MSW compost having much lower organic matter (<14% C) has the potential to fit into the INM in crop production and can substitute considerable fraction of fertilizer nutrients.

Matter of Concern

Heavy metal content in the rural composts is quite low as it is prepared from crop residues and cattle-shed litters. Hence, though these materials are being used by farmers of the country in their land since many years ago, build-up of metals in soil and contamination of crop were never matters of concern. Due to increased use of metals and different synthetic chemicals in changing lifestyle of urban residents, toxic pollutant concentrations has increased in the MSW, resulting in enhanced threat to all components of the environment. A survey in composition of composts prepared from such wastes in 29 cities indicated (Saha et al., 2010) that heavy metal contents varied widely. The Fertilizer (Control) Order 1985 (FAI, 2007) has defined quality control (QC) protocol for the purpose of regulating the production and marketing of high quality city waste composts in India. Only five MSW compost samples analyzed contained all the heavy metals within the above QC limit and all of these had been prepared from segregated wastes. Other samples had 1, 2, 3 or 4 heavy metals (out of the 6 analyzed) more than this limiting values and degree of contamination varied widely. MSW composts produced in bigger cities (>1 million population) had, on average, 122% higher concentration of total heavy metals compared to those prepared in smaller cities (Figure 1).

Composts prepared from segregated wastes contained on an average 68% less heavy metals as compared those prepared from non-segregated wastes (Figure 2). Partial segregation of wastes before composting also reduced heavy metal load in the composts by 33% on average. In some cities, viz., Composts prepared from decomposable wastes which were segregated at household level contained significantly lower content of heavy metals, by about 51% as compared to composts prepared from wastes which had been collected from places known to generate mainly biodegradable wastes like, vegetable market, hotel & restaurants, slaughter house etc. This indicates that segregating of biodegradable wastes at household level can reduce metal contents in the compost to a minimum.

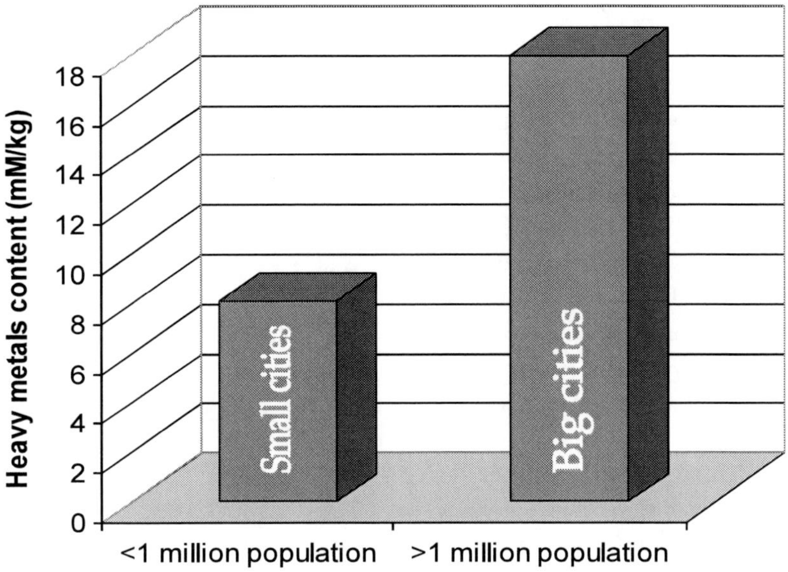

Figure 1. Average heavy metals content in MSW composts prepared by Indian cities with different size.

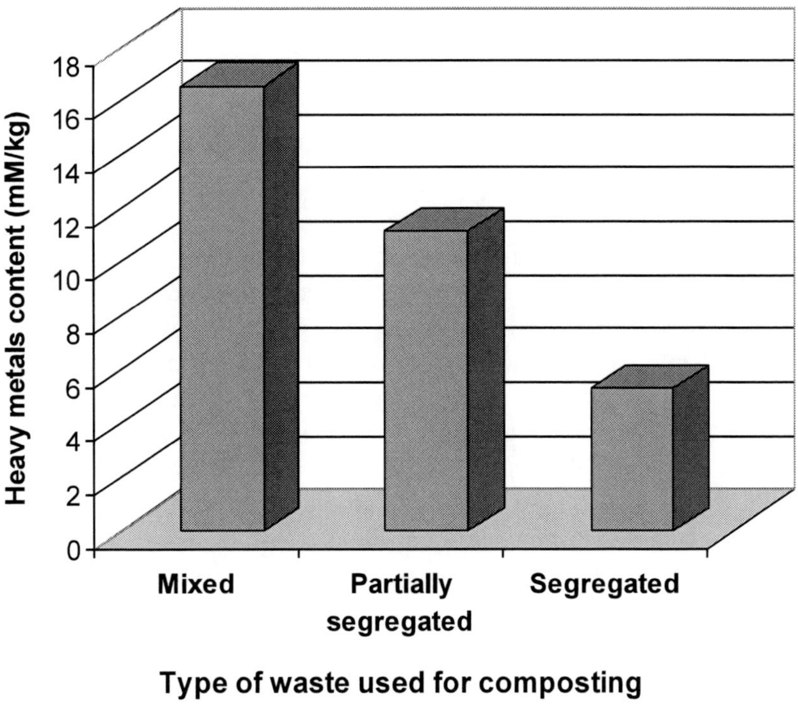

Figure 2. Effect of segregation of wastes prior composting on heavy metals content in the MSW composts.

SYNCHRONIZATION BETWEEN SWM AND INM

As different methods are followed for composting of MSW in Indian cities, quality of resultant organic manure varies widely in respect of nutrient supplying potential as well as ability to pollute land. Authors computed 'Fertilizing index (FI)' (an indicator of potential for improving soil fertility) and 'Clean index (CI)' (an indicator of low content of toxic heavy metals) for the MSW composts produced in 29 cities of India (Saha et al., 2010). Fertilizing index value varied widely from 1.79 to 4.16 with a mean value 2.95 and relative standard deviation (RSD) of 22.5%. Fertilizing index values of cattle dung manure (CDM) varied from 3.7 to 4.06 with mean value 4.12. More than 90% of MSW compost had fertilizing index values less than those found for CDM. Further composts produced from mixed wastes had on average 20% lower Fertilizing index compared to those produced segregated biowaste compost. While CDM had maximum CI values of 5, MSW composts had CI values ranging between 0.47 to 5 with a mean value of 3.12 and RSD of 35%. Composts produced from mixed wastes had on average 50% lower CI compared to those produced segregated biowaste compost. Further, the composts produced by three manufacturers from household level source-separated biowaste compost had CI values equal to that of CDM. On the basis of FI and CI values, Saha et al. (2010) categorized MSW composts into two major categories: suitable or unsuitable for land application (Table 8). Those suitable for land application were further categorized into marketable (A, B, C and D) and restricted use classes (RU-1, RU-2 and RU-3). Results showed that majority of the MSW compost samples (69%) were unsuitable for land application either due to very low nutrient supplying potential and/or high heavy metals content. Among the samples suitable for land application, only 27% belonged to marketable classes (A and B) complying with regulatory limits of India for heavy metal contents.

Although there is no recommended upper limit for application rate of compost prepared from rural wastes, it generally varies between 5 and 10 t/ha of land as part of INM (along with fertilizer nutrients) for growing field crops (ICAR, 2005). On the contrary, average availability of dung manure for land application in almost all the states is less than 3 t/ha (IISS, 2010). In order to supplement the deficit of organic manure, compost prepared from MSW has been indicated as second largest source manure to fit into the INM. However, high concentration of heavy metals is seen as an obstacle for its application in agricultural land. For example, organic farming requires compost with very high nutrient supplying potential and very low content of toxic metals in order to ensure high yield as well as quality of the produce. Vegetable crops, in general, have high capacity of metal uptake from soil and therefore, INM should also focus on restricting concomitant metal entry through compost. A number of non-food crops/plantation, like jute, cotton, castor, linseed, rubber etc. are grown in the country, where plant uptake of metals may not be a great concern. Mode and frequency of application of compost also vary widely with nature of crops/trees. While most of the food crops and some commercial crops require frequent application (generally annually) on the upper 20 cm deep layer of land, establishment of garden, lawn, forestry and fruit trees require deep or surface incorporation of heavy quantities of manure at the time of establishment.

Suitability of MSW Compost Produced through Different Methods

Compost prepared from mixed MSW: About 39% of manufacturers manufactured compost from MSW which were not segregated at all before composting. Most of the manufacturers who installed mechanical composting plant for handling large volume of wastes were producing compost in this way. About 79% of the composts produced from mixed wastes were unsuitable for land application and remaining (under RU-3) were suitable only for one time application in a land for developing lawns, gardens, establishing vegetation/forestry etc.

Table 8. MSW compost samples produced in India belonging to different categories

<table>
<tr><th colspan="3">Categories of MSW compost</th><th>% sample in the category</th><th>Probable use in different category of land use</th></tr>
<tr><td rowspan="7">Suitable for land application</td><td rowspan="4">Marketable classes</td><td>A</td><td>5.5</td><td>Best quality.
High Manurial value potential and low heavy metal content and can be used for high value crops, like in organic farming</td></tr>
<tr><td>B</td><td>2.8</td><td>Very good quality.
Medium fertilizing potential and low heavy metal content</td></tr>
<tr><td>C</td><td>-</td><td>Good quality.
High fertilizing potential and medium heavy metal content.</td></tr>
<tr><td>D</td><td>-</td><td>Medium quality.
Medium fertilizing potential and medium heavy metal content.</td></tr>
<tr><td rowspan="3">Restricted use classes (should not be allowed to market)</td><td>RU-1</td><td>5.5</td><td>Low fertilizing potential. However, these can be used as soil conditioner.</td></tr>
<tr><td>RU-2</td><td>5.5</td><td>Can be used for growing non-food crops.
Requires periodic monitoring of soil quality if used repeatedly.</td></tr>
<tr><td>RU-3</td><td>25.0</td><td>Can be used only for developing lawns/ gardens (with single application).</td></tr>
<tr><td colspan="3">Unsuitable for agricultural land application</td><td>55.7</td><td>Can be used for rehabilitation of degraded land.</td></tr>
</table>

Compost prepared from partially segregated MSW: One-third of the MSW compost manufacturers made efforts to remove large non-biodegradable fractions from the mixed MSW received at the composting ground. In some cases, composting was done using MSW with lower density (assuming biodegradable fraction being lighter). About 50% of the composts produced from mixed wastes were unsuitable for land application, 41% belonged to restricted use classes (8% in RU-1 and 33% in RU-3) and remaining were classified under marketable class D. Variability in feedstock mixture and degree of segregation has caused in such a large variation in grades of composts prepared from partially segregated MSW..

Compost prepared from segregated MSW: About 28% of manufacturers prepared compost from segregated biodegradable fraction of MSW. Biodegradable wastes are segregated at generator level in two ways. In few cities (Namakkal, Suryapet and Vijaywada), individual households deliver segregated biodegradable wastes separately during 'door-to-door' collection by municipal organization, which are composted in pits or heaps using earthworms. In other cities, sources generating mainly biodegradable wastes, like hotels & restaurant, vegetable market, slaughterhouse etc. are selected for feedstock collection for composting plants. Seventy percent of the composts produced from segregated bio-wastes were suitable for land application. Half of the samples under this category contained heavy metals with the regulatory limits and were suitable land application for growing any kind of crops. The composts produced only from source separated biowastes following 'door-to-door' collection had high manurial value and were categorized under either 'A' or 'B' classes. During the composting process (also known as vermicomposting) followed for all of these three composts, worms were used mainly for hastening the decomposition process and materials were handled by labours using small implements. Compared to cattle dung manure, such composts had considerably higher N content, but similar content of P, K and organic matter.

CONCLUSION

Large volume of solid wastes generated by ever increasing urban population of India demands their disposal in environmentally safe manner. Composting biodegradable fraction of the waste to prepare a useful product for agriculture sector has been found an effective way of reducing pollution burden on the environment. At present, volume of recycling of wastes in this manner is quite small. Although, mechanical compost plant had been installed in several cities (22%, Kumar et al., 2009) to handle large volume of wastes processing, none of these are producing composts suitable for agricultural crops. The main reason for producing poor quality compost is the use of mixed MSW as input material for composting. Results clearly indicated that composts produced using mixed MSW had low organic matter, plant nutrients and high toxic metal contents. Removal of large non-biodegradable fraction from mixed wastes prior composting did not have significant impact of compost quality to be eligible for use in INM for crop production. Among the composts prepared from biodegradable fraction of MSW, only those prepared using source separated biodegradable wastes belonged to marketable grade and suitable for producing food crops. Based on analysis of information generated in the field, following recommendations can be considered for implementation.

(a) Grading of composts prepared from MSW on the basis of manurial value and heavy metal contents is expected to maximize recycling of MSW by suggesting alternative use of poor quality of composts, while protecting land resources from pollution.
(b) Biodegradable fraction needs to be separated out from the solid wastes at the source of generation itself before composting. For this, continuous effort should be made to

create awareness among the citizen for storing biodegradable and non-biodegradable wastes separately and among the local public bodies for collection of separated wastes from door-to-door and transportation of biodegradable wastes to composting yard.

(c) High cost of transportation and large capital cost for setting-up of mechanical compost plant are two major factors responsible for high manufacturing cost of compost. Setting up small composting yard at ward level can minimize cost of transportation and use of manually operated smaller implements for handling of wastes during composting can minimize cost of compost production further.

(d) Sewage sludge should not be added at any point of time since it will raise the metal content of the compost.

(e) Nutrient availability, specifically nitrogen content varies widely among the composts prepared from segregated biowastes probably due to variation in composting methods. Optimization of the compost production parameters that increase nutrient availability need to be identified and widely used in MSW composting.

(f) Availability of phosphorus is generally low in Indian soils. Also, MSW composts are poor in phosphorus content irrespective of method of composting. Organic acids produced during decomposition of organic matter have the ability to solubilize insoluble inorganic phosphorus compounds. With this principle, a technology was developed to prepare 'phosphocompost' by mixing ground rock phosphate with rural organic wastes before composting for the purpose of increasing content as well as availability of phosphorus (Singh and Tripathi, 2000). This technology can also be applied for enhancing the fertilizing potential of MSW composts.

REFERENCES

Bhattacharyya, P., Chakraborty, A., Bhattacharya, B. & Chakrabarti, K. (2003). Evaluation of MSW compost as a component of integrated nutrient management in wetland rice. *Compost Science & Utilization* 11, 343-350.

CPCB (2005). *Municipal Solid Waste: Implementation of MSW (M&H) Rules.* Central Pollution Control Board, Ministry of Environmental and Forests, Governments of India, New Delhi. http://cpcb.nic.in/oldwebsite/Municipal Solid Waste/Implementation_of_ MSW_ M&H)_Rules. html (accessed 13.08.2008).

FAI (2007). *The Fertiliser (Control) Order 1985.* The Fertiliser Association of India, 10, Shaheed Jit Singh Marg, New Delhi, India.

FAI (2010). *Fertilizer Statistics 2009 – 2010.* The Fertilizer Association of India, 10, Shaheed Jit Singh Marg, New Delhi, India.

Hargreaves, J.C., Adl, M.S. & Warman, P.R. (2008). A review of the use of composted municipal solid waste in agriculture. *Agriculture, Ecosystem and Environment* 123, 1-14.

Howard, A. (1943). *Agricultural Testament.* Oxford University PressInc., London.

Kavitha, R. & Subramanian, P. (2007). *Effect of enriched municipal solid waste compost application on growth, plant nutrient uptake and yield of rice.* Journal of Agronomy 6: 586-592.

Khiani, K.N. & More, D.A. (1984). Long-term effects of tillage operations and farmyard manure application on soil properties and crop yield in a vertisol. *Journal of Indian Society of Soil Science* 32: 392-393.

Kulkarni, K.R., Hukeri, S.B. & Sharma, O.P. (1978). *Proc. India/FAO/Norway Seminar on Development of Complimentary use of Mineral Fertilizers & Organic materials.* Ministry of Agriculture & Cooperation, New Delhi.

Kumar, S., Bhattacharyya, J.K., Vaidya, A.N., Chakrabarti, T., Devotta, S. & Akolkar, A.B. (2009). Assessment of the status of municipal solid waste management in metro cities, state capitals, class I cities, and class II towns in India: An insight. *Waste Management* 29, 883 – 895.

Lal, S. & Mathur, B.S. (1989). Effect of long-term application of manure and fertilizers on the DTPA-extractable micronutrients in acid soil. *Journal of Indian Society of Soil Science* 37, 588-590.

MOEF (2010). *India: Greenhouse Gas Emissions 2007. INCCA Indian Network for Climate Change Assessment.* Ministry of Environment and Forests, Government of India, New Delhi.

MOUD (2005). *Report on 'Management of Solid Waste in Indian Cities'.* Ministry of Urban Development, Government of India, New Delhi.

NEERI (1995). *Strategy paper on SWM in India.* National Environmental Engineering Research Institute, Nagpur, India.

Rabindra, B. & Gowda, H. (1986). Long range effect of fertilizer, lime and manures on soil fertility and sugarcane yield on a red sandy loam soil (Udic Haplustalf). *Journal of Indian Society of Soil Science* 34, 200-202.

Saha J.K. (2003). *Completion Report of Research Project 'Nutrient management practices for important oilseed based cropping systems for improving yield and oil output under rainfed conditions'.* Indian Institute of Soil Science, Bhopal, India. pp. 44.

Saha, J.K., Panwar, N.R., Singh, M.V. & Subba Rao, A. (2008). *Quality indices of municipal solid waste compost for safe use in agriculture.* Technical Bulletin No. 6. Indian Institute of Soil Science, Bhopal, India, pp. 44.

Saha, J.K., Panwar, N. & Singh, M.V. (2010). An assessment of municipal solid waste compost quality produced in different cities of India in the perspective of developing quality control indices. *Waste Management* 30, 192–201.

Sarkar, A.K., Mathur, B.S., Lal, S. & Singh, K.P. (1989). Long-term effects of manure and fertilizers on important cropping systems in sub-humid, red and laterite soils. *Fertilizer News* 34: 71-80.

Singh, M. & Tripathi, A.K. (2000). *Phosphocompost: A complete natural manure.* Technical Bulletin. Indian Institute of Soil Science, Bhopal, India. pp. 13.

Swarup, A. (1998). *Emerging soil fertility management issues for sustainable crop productivity in irrigated systems.* p.54-68. In: A. Swarup, D. Damodar Reddy and R.N. Prasad (eds.) Proc. National Workshop a Long-term Soil Fertility Management through Integrated Plant National Supply Systems, Indian Institute of Soil Science, Bhopal, India.

In: Municipal Solid Waste
Editors: G. Konno and C. Machado
ISBN 978-1-62100-257-4

Chapter 3

ISSUE OF MUNICIPAL SOLID WASTE AND THE PROCESS OF ITS MANAGEMENT IN SOLUTION

Tuba Öztürk[a*], Sevil Veli[b] and Anatoly Dimoglo[c]

[a]Department of Environmental Engineering, University of Namık Kemal, Corlu, Tekirdag, Turkey

[b]Department of Environmental Engineering, University of Kocaeli, Kocaeli, Turkey

[c]Department of Environmental Engineering, Gebze Institute of Technology, Gebze, Kocaeli, Turkey

ABSTRACT

Municipal solid wastes are accepted less harmful than the other kind of wastes in terms of environmental and public health depending on the process of their occurrence. The wastes determined as municipal wastes involve malfunctioned forms of products such as organic substance, plastic, textile, paper, wood, metal and glass. The proportion of these substances is being changing in accordance with the cultural structure and life standards of societies. In the world, there are various approaches for controlling municipal solid wastes. Yet the methods used for the solution of the problem have been insufficient in their own. The gretest problem arising from solid waste is primarily the increasing amount of these wastes with each passing day. The focus of the studies performed recently on this issue is to immobilize the amount of waste or at least slow down the rate of increase. However, nowadays, it is clear that the situation will conflict with the economic policies supporting the consumption growth and needs long-termed studies.

For waste management the necessity of building on the acceptance of waste raw materials by taking into account the durability of needs and limitedness of natural sources is obvious. Generally, a successful waste management should keep various eliminating methods together and be supported with laws in order to get ahead with the methods applied. The significant point in here, and particularly within the scope of waste-raw material relation, is that any society being in the position of producer should be a part of the process of waste management and be encouraged to increase efficiency of recycling. As well as increasing the success

* Corresponding author: Tel: +90 282 652 94 75; Fax: +90 282 652 93 72; E-mail: tubaozturk@nku.edu.tr.

of recycling processes by allocating responsibility to its producer with regard to recognizing and separating its waste, the control of hazardous wastes arising from domestic use also gains importance. Thus, enabling municipal wastes to be in less harmful position will lead to progress in the context of the environment protection in a short time.

Keywords: Municipal Solid Waste, Waste Management, Recycle, Waste Producer

1. Introduction

Production and consumption, briefly the extent of pollution arising from daily life, indicate that environmental issues are actually global problems. Owing to its being global, the problem reveals the reality about the requirement of methods used in its solution should also be global. The concept lies behind all issues related to pollution is waste.

Waste is any sort of substances generally unnecessary, nonwanted, having no market with its generated form and involving removal safely [1]. These substances are also known as rubbish, garbage, trash, refuse besides being called as solid waste [2].Wastes are not only generated by human but also they are arisen from plants and animals [3].However, it is linked to the waste originating from human being that makes it an environmental issue. For that reason, it is accepted that waste is generated during the processing of materials or the consumption of final products that are used or needed by people for their daily lives [4].

The first waste is usually generated within the period starting from the processing of raw materials into intermediate and final products to the consumption of final products by the consumer. The waste generated by the consumer is the last waste. Wastes are found in forms of solid, liquid and gaseous according to their generation [1]. The fundamantal point while characterizing wastes in this manner is the overall situation of waste. Therefore, it is sufficient for any wastes not to contain liquid in terms of being fluid in order to be defined as solid waste. Solid wastes can also be classified in their group in various ways with regards to their features and resources. The most common classification about this matter is to categorize solid wastes as municipal and hazardous solid wastes.

Municipal solid waste is a term mostly used to refer heterogeneous mixture of the wastes generated from residential areas.Depending on the characteristics of a city, the composition of municipal solid waste differs from region to region. This waste is basically includes miscellaneous waste from residential, commercial, various institutional sources, trashes from parks and yards and excavations from construction and demolition. The wastes generated in these sources are predominantly food, paper, textile, ashes, plastic, glass, metal and wood [5-8]. The amount and feature of wastes generated in a region link to the living standards and lifestyles of the local people. Besides this, climate, sources and sustainability of the sources of the region have an effect on wastes. The main source of municipal solid waste flowing is residences correspondingly domestic solid wastes. For that reason, it is common to use the notion of domestic solid waste instead of municipal solid waste.

The term municipal solid waste is mostly used to state the category accepted as nonhazardous.

However, wastes, having at least one of the qualities of being ignitable, toxic, corrosive and reactional that are defined as hazardous features, have especially participated municipal solid waste flow recently. The main source of these substances are hobby chemicals coming from houses, paints, cleaners, oils, batteries, pesticides. These wastes arising from residential sources are defined as domestic hazardous waste [3,9,10].

Apart from these wastes, namely, municipal non-hazardous solid wastes have subdivisions as organic and inorganic solid wastes. Also, organic solid wastes can be classified in itself under three main headings as being putrescible, fermentable and non-fermentable wastes. Putrescible wastes are the kitchen wastes decomposing fast without requiring to be controlled and giving rise to outlook problem; fermentable wastes includes the wastes decomposing fast without requiring to be controlled and not having an odour problem; non-fermetable wastes cover the wastes resisting against decomposition, therefore, they consist of wastes decomposing very slowly [5].

Solid wastes interact with the environment and people either directly or indirectly from the moment they are generated until the phase of disposal. For that reason, they have a great deal of risk in terms of enviromental and public health and their control is very important. Depending upon increasing and differentiating needs all over the world, the amount of generated solid waste is becoming more and more. Therefore, it has been gone through a municipal solid waste problem arising from the amount and composition of waste both in developed and developing countries. All countries in the world have serious worries about the present state and future outcomes of the problem [3,11-14].

2. Solid Waste Management

The concept of solid waste management in developed countries has been resulted from owing to the fact that it has been gone through such a large scale. The waste management covers the activities that minimize the negative effects on environment and health of the wastes mentioned [2]. Each kind of project improved for that purpose, applicable program and alternatives make up the components of solid waste management [15].

In parallel with rapidly growing solid waste generation, the difficulties encountered its disposal are the common issues of all the countries around the world. It is necessary for the countries suffering from this problem to cooperate and share knowledge and experience in order to get succeed in solid waste management systems. In general, solid waste management is a whole process of collecting, carrying and disposing of waste in a manner without threating the environment and public health [16].

The recognization of solid wastes as an environmental issue started with the rapid growth of the amount of generated waste.The disposal of growing waste masses started to give rise to both economic and health problems, but only after that it was recognized as a problem and brought to the agendas of governments.

Neither the solid waste problem nor the solid waste management and system of hierarchy approaches developed for the solution are new for today. However, this management approach has been limited to even in developed countries with regards to manicipal solid waste collection, storage and incineration in landfill for a long time. The start of associating health problems experienced in these countries with environmental pollution over time led to the public recognization of the outcomes of pollution by giving rise to create an essential environmental awareness. Due to the increment of public interest and awareness about environmental issues and the progresses in environmental science and technologies have started to push the authorities about the soulutions of the problems [17,18]. In addition to all these problems, the inefficiency about the methods of incineration and storage on their own related to generated solid waste masses has lead to a sustianable improvement with a more modern approach of solid waste management [17].

3. Sustainable and Integrated Solid Waste Management

Sustainable and integrated solid waste management provide a more systematic view to the problem [2]. As a whole, this approach aims to ensure reducing waste amount in its sources, proper separation, collection, transportation for generated wastes, and recycling materials at a maximum rate from the volume of it, disposing the remains in accord with the principles of public health, economy, engineering, aesthetics and other environmental conditions [16,19].

Sustainable and integrated solid waste management is based on a safe and effective waste management disciplines. Its starting point is the limitation of raw materials and resources. The approach has embraced the control and disposal methods such as waste minimization, collection, transportation, final disposal of basic waste management. However, incineration and storage methods for disposal of large volumes waste for a management system aren't found sufficient [17,20]. Mostly, it is a widespread belief that applied solutions have given rise to further growth of the problem. It also emphasis that these procedures should be managed intercorelatedly with the principle of consistency and complementarity. Sustainability management system approach is based on the adaptation of national or local extent. Therefore, environmental management plans that are formulated at the national level is the basis of the system. Waste management is both an integrated form of multisectoral planning and a long-term process [3,21].

Sustainable and integrated solid waste management is the component of sustainable development desired to be accompolished. This approach needs to be supported by techonology during its complementary stages such as public awareness, economic and financial capasity as well as the stages involving environment, public health and management of natural sources. All organizational sturctures required continuity of the process and the establishment of the managment system, appropriate legal and framework regulations are necessary for the achievement of the system [21].

These waste management hierarchy can be applied to all waste types. On the other hand, the reguations about collection and disposal of medical wastes particularly in developing world cities are not at a sufficient level. The hierarchical structure with regards to waste reduction, recycling, incineration and storage requires the optimization of the system management with a view of critical examination.

It is mostly under the responsibilities of local government to manage the flow of solid waste [2].Municipal governments should be able to provide adaptation for the combination of various projects and techniques to the area [15]. Sustainable and integrated solid waste management is an integral part of a good comprehension of local government. Because the service area has the easiest effect on the perception local governments. Waste management is a versatile programming. Besides its technical, economic and socio-cultural content, it has also contents of national and global policies and principles in this regard [2]. Being so extensive and ambitious, the system requires wide range of participation and sharing responsibilities, as well. That can be only ensured with a team approach encouraging all participants.

The members of the system are the main participants of the system that is system administrators, local authorities, central government, private sector, all non-governmental organizations and public [21]. Solid waste management being one of the most important headings of environmental management in our day is open to all parts serving the target of the system. It is a must to be in contact and cooperation especially with the residents of the area getting service for the success of the system [2]. A sustainable and integrated solid waste management should be ruled out based on transparency, equality and the principle of accountability. An effective waste management can be only achieved with an estimation of all components and all factors having an effective on succeed. Besides this, targets should be determined perceptibily. In this regard, setting realistic goals, posing appropriate programmes and defining each step on account of reaching the goal are essential. Sustainable and integrated solid waste management is a long-term process. Therefore, one of its most important pillars is the economy. It is necessary to put aside the financial sources for the applicability of the program in the long run [15]. A modern solid waste management program should be applied to a reasonable cost. This is important, while the management systems imposed mostly in developing countries have high cost, the level of service received, namely, the success of the system is low. In such cases, a comprehensive analysis of cost and expenses with a rational replanning is necessary for the continuation of the system [2].

The system can be developed to solve the problem at country, region, state, city extent [19]. The important thing in here is to have been formulated the detailed principles and the process of the system. Sustainable and integrated solid waste management covers to minimize resource use and its environmental impacts. It aims to manage the environment carefully by protecting natural sources and providing the recycling of wastes at convenient places [2]. The initial point is to enable the reduction of generated waste as much as possible and consider generated wastes as sources.

3R approach with its positive environmental impact has a significant place for sustainable development approach. The solid waste management approach of The Environmental Protection Agency is stated with 3R, a three-phase approach incuding reduction, reuse and recovery. Reduction is defined as minimizing waste generation; reuse is utulizing waste substances again and again; recovery is described as transforming wastes to be able to be resused [22].

The success of sustainable and integrated solid waste management is compelling because of the main principles that is based on and its goals for the solution of solid waste problem. No country in the world in terms of solid waste concepts is away from the problem and solution. As well as environment and public health, there is no other chance for all countries in the world about being successful and applying this management system properly in order to protect natural sources.

3.1. Waste Characterization

The strategies performed during sustainable and integrated solid waste management process are determined depending on the amounts of wastes besides their physical and chemical features [19]. It is the first step of a successful management the adequete prediction of the amount of waste generated in a residential area. It is known that countries with low income have 0,3 -0,5 kg / daily amount of waste per capita, whereas countries with high income have 1,6-2 kg/daily per capita. It should be gained adequate data about the source and composition of wastes besides their amounts in waste management [12,15]. Researches have shown that generated municipal solid wastes on a large scale account for paper, plastic, glass and metal containers with low density in industrial countries while they account for sand, ashes, rock and dust containing inert materials with high density in developing countries. Besides these, the consumption of fruit and vegetables in high quantity depending upon nutritional habits of societies also increase the moisture content of wastes correspondingly their density [2].

These data in any dwellings or towns are effective all along the line about planning the collecting system ranging from features of tool to making a decision of collecting frequency [15]. Well-operated system and vehicles, coming from industrial countries, with low density wastes are not safe and suitable for high density wastes. The content and weight of wastes cause break down of equipments rapidly due to corrosion [2]. Knowing physical and chemical features of wastes is the most effective element in the selection of appropriate recycling method [15]. Since the further content of wastes with regards to moisture and inert material, it means that they are not suitable for incineration. Apart from this, wastes in the case of containing great amount of organic substance requires composting, on the other hand, they necessitate to work on other recycling processes in the event that they involve materials with economic value. Certain chemicals in waste should be at certain levels for composting. Deciding the application of co-composting for nutrient supplement in case of their deficiency is one one of the issues having to be planned initially for a sustainable management [2].

3.2. Waste Minimization

The starting course strategy in solid waste hierarchy is the reduction of waste generation arising form both industrial production processes and household use. Reduction of wastes in resource leads to the improvement of clean technologies and the usage of processes generating less waste. However, it takes along time and effort to develop new technologies or the adjusment of processes in use. This step in waste management also covers betterment, producing foolproof products and enabling to be preferred by consumers for use. It aims to limit the usage of hazardous substance during production process and avoid using it completely over time, too. As well as providing substantial amount of raw-material and energy conservation, it is a particular process that lowers the cost of decomposing [23-25]. The most important feature of this process is the requirement to count consumers in charge apart from producers. Hence, there is a great need for comprehensive training and field work.

3.3. Collection and Transportation of Waste

Waste collection process covers taking solid wastes from their resources to transpose them to transfer stations and then being carried to final disposal points by large container trucks, rolling stocks or large transport vehicles like barges in accordance with hierarchy of the applied system.

Collection of municipal solid wastes is a process that requires continuity at regular intervals. Collection process can only reach its goal provided that it is performed in optimum frequency.

Various factors particularly needs, desires, health of local people and others like workforce, cost are effective to detrmine the frequency of waste collection. Collecting wastes rarely means the increment of duration about keeping wastes at homes, on the other hand, it accounts for the decrease in the number of workers and cost of fuel. Although this circumstance provides advantage in terms of cost, it gives rise to be questioned with regards to the service quality of management system. In this respect, it is a must for health and aesthetic to collect wastes at least once a week by considering vehicular traffic.

The best part of total cost involves cost of collection and transport in solid waste management systems. According to this, the success or failure of most solid wastes management systems depends on the success or failure of collection system [15]. For that reason, it is almost a prior condition to for the success of sustainable and integrated sold waste management to have a planned urbanization and infrastructure.

The collection of municipal solid wastes are under the responsibilities of local governments in cities. The process of collecting wastes for some cities in developed countries is performed by private companies getting the tender on behalf of municipalities. Not being collected solid wastes properly and their insuffient disposal are some of the biggest problems in big cities they faced [26]. These cities experince severly all of the problems of fast and nonfoundational urbanization. Rapid population growth gives rise to exceeding the capacity of infrastrucure in a short time and for local people not getting basic service from local governments. In general, solid waste collection plans in these cities consist of only limited part of total city population [2].

This case accounts for the solution of the problem in illegal ways for the population of low income being out of plan. Noncollectible wastes in suborbs are left roadsides, canals, free fields [27]. Wastes being spoiled, stunk over time pave the way for the development of various diseases and their spread. For that reason, being able to talk about waste management in a settlement requires to perform primarily collecting service permenantly and healthily. Hence, it is very important to be careful about the selection of equipments, especially ensuring the adjustment of imported technologies to local conditions, to prefer simple and workforce oriented management systems as much as possible. Besides this, worker recruitment and training in workforce oriented strategies is one of the critical issues for management system [21].

3.4. Recycling

Recycling is the transformation of reusable substances by applying various physical and chemical process so as to be turned them back to the market in solid waste flowing [28]. It is very important for sustainable development to recycle or reuse substances in solid waste and it has an important place in the management hierarchy [29-31]. Its success means that the whole system reaches its goal.

Being put into reproduction process by coverting into sub-raw material, it enables to the use of wastes as resources and the protection of natural sources. Besides this, preventing pollution which might be generated during extracting and reproducing, enabling energy saving from the energy being likely to used in the course of these processes, increasing employment opportunities, reducing workload of the step following itself in hierarchy are the most contributions of recycling to the modern world [30]. Depending upon the success of recycling, extending the operating time of storage areas, enabling the separation of materials such as glass and metal from waste flowing in terms of affecting positively the processes of both the amount of ashes generated in incineration plants and the process of composting are considered as its indirect advantages [28,29].

Although recycling is accepted as one of the most important steps of managemnet system, it has mostly been developed by illegal sectoral activities. It is pointed out that the main reason for that is based on the inexpensive and heavy workforce.

In addition to the development of a sustainable recovery and recycling products, demand for the creation of new technologies is a must [29]. In general, among low-income people in developing countries the re-use of waste materials is a common practice. In other words, recycling is a method based on very old foundation of individual and familial level. Finding a place in waste management system and becoming more commercialised, recycling is the result of the supported studies related to this issue performed all over the world. In particular non-governmental organizations in developing countries have a growing interest and support efforts for recycling [21].

Recycling process, with a special place in sustainable and integrated solid waste management, is a permanent potential market for recycled materials, the most important factor, in terms of being successful and permanent. The reality of constant market is the constitution of consumer group wanting to buy the products generated by secondary raw-material. Apart from the requirment for the continuity of product, this case has great importance with regards to enabling the growing and improving reliance on recycling [28].

Municipal solid wastes generated in the areas of settlements being socioeconomically powerful have high rate rcycling potent. They have a recycling mechanism that is quite developed and established for some materials particularly aliminium and glass. The main reason for it is to have been built a improved demand, namely, a market based on areas of usage and properties of these materials. On account of the development of recycling system completely, there is a need for building up a serious control mechanism checking product quality as well as supply and demand. It is known to have experienced market losses depending on especially the alteration of paper quality in the past. For that reason, it should be kept in mind that recycling can only develop as long as it is performed properly and appropriately.

Besides this, it should be remembered that recycling is not always the most suitable choice. It shouldn't be preferred as a method on condition that a recycling material contains much raw-material, or recycling method isn't economic, or it causes either high energy consumption or pollution [23].

What is important in waste management system is that recyccling prrocesses don't give rise to environmental pollution with regards to its purpose and starting point. Managing toxic substances, revealed during these processes, without leading to any pollution is a matter that must be dealt with carefully especially in developing countries. Recycling processes performed illegaly in these countries not only leads to health problems for the people working in this field but also causes environmental pollution at various sizes [28].

3.5. Composting

Composting is the biochemical separation of organic compounds of solid wastes. The various microorganisms found naturally in solid waste during composting lead to the decomposition of organic substance under controlled conditions. As a result of composting, a product recovering the structure and the quality of soil is obtained. With a past dating back to old times, composting method is significant with regards to transforming particularly kitchen, park and yard wastes into useful product [32-35].

A number of techniques are defined in literature for composting process performed under aerobic or anaerobic conditions. The process pace of aerobic composting methods have some superiorities such as causing no problem related to any odour and pathogenic microorganisms [15,23].

The success of waste composting process depends directly on the preparation and process control. It is obvious that the composting products placed wrongly and prepared disagreeably will give rise to soil and groundwater pollution [28]. Composting is an important recycling process in waste management system. Furthermore, it can be operated with all other recycling processes in parallel with the complementarity principle of management system. The need for composting as a recycling process in municipal solid waste management over time is apparent [18].

3.6. Incineration

Incineration is an important method used in solid waste disposal for many years. When performed in well-designed plants, it is not only the most efficient volume reduction method but also it enables the control of medical and hazardous wastes [8]. In general, when wastes with high calorific values are burnt, enrgy creation is at stake. The energy received can be both turned out electrical energy and utulized in heating of large buildings. Incineration process can be evaluated as a recycling process on condition that it provides not only the reduction of waste amount but the generation of heat and electrical energy as well [36]. However, as well as having these advantages, incineration, known as waste disposal procedure in managment system owing to with its high investment cost, complex control emmision, is a rather expensive and high-powered method. Since creating awareness in public due to ashes made up after the process and especially flue gas emmisions, it must be monitored carefully by the system administrators [23,28].

3.7. Storage

The most preferable disposal method in many parts of the world about solid waste management is storage [31]. Approximately 70% of all waste produced in the world are disposed by using the storage method [37,38].Storage is a method required to be followed in terms of environment and public health due to leachates and landfill gases making up in these areas. Performed in areas, especially in developing countries, where the process of storage is uncontrolled and lack of leaking is a potential danger [39]. The storage areas being out of the cities previously are rapidly surrounded by dwellings owing to uncontrolled rapid growth of towns in these countries. Residential and solid waste storage areas with time have started to threat public health directly. That's why, storage process is only accepted as a stage provided that it is performed in controlled areas designed properly [40].

Storage method involves the final procedure in the hierarchy of waste management storage method.Considering the troubles experienced about the storage areas being packed rapidly and opening new storage areas, the acomplishment of former administrations have become more and more crucial.

Storage areas for feasibility of the method is of great importance for the site selection. Establishing storage areas away from residential areas means high cost transportation as well as accepting financial burden of infrasture investments like roads [2,23].

CONCLUSION

Nowadays nearly all habits concerning modern life are consumption-oriented and particularly consumption has taken the shape of a lifestyle in the societies being powerful socioeconomically. Modern culture and lasting needs based on consumption give rise to versatile environmental impacts. These environmental impacts have already crossed the lines that nature can tolerate. That's why, environmental pollution has been perceived easily by anyone and experienced in an extent that can be established a connection in terms of cause-effect relation. Solid wastes are one of the most significant reflections of the economic structure constructed in line with industrilization and consumption in our century. Although the generation of solid waste requires a socioeconomic base, it has an impact on all countries in the world essentially the developed ones.

The rise of solid waste around the world correspondingly the amount of waste in each day accounts for being short of the methods used to control these wastes. The failure about eredicating the problem and the obstacles experienced during the attempts of finding solutions require to share different approaches, knowledge and experience. The issue of solid waste should also be evaluated as problems of source and raw-material. However, it should be take into account that the approach will require close collaboration between countries and acquire global perspective in terms of utilizing world sources and sharing responsibilities. It has great importance for a sustainable and integrated waste management to aquire success in terms of putting into practice with a view of concerning criticism, development, open to new ideas.

ACKNOWLEDGMENTS

The authors express their thanks to Ayça Özgüven for their help.

REFERENCES

[1] Igoni, A.H.; Ayotamuno, M.J.; Ogaji, S.O.T.; Probert, S.D. *Appl Energ,* 2007, *vol 84,* 664-670.

[2] Zhu, D.; Asnani, P.U.; Zurbrügg, C.; Anapolsky, S.; Mani, S. *Improving Municipal Solid Waste Management in India;*ISBN-978-0-8213-7361-3, The World Bank, Washington, USA, 2008, Vol.1, pp. 1-8.

[3] Sasikumar, K.; Krishna, S.G. *Solid Waste Management;* ISBN-978-81-203-3869-2, PHI Learning Private, Eastern Economy Edition, New-Delhi, India, 2009, Vol. 1, pp.1-33.

[4] Tchobanoglous, G.; Theise, H.; Vigil S. *Integrated Solid Waste Management, Engineering Principles and Management Issues;* Mc-Graw-Hill Inc., Sıngapore, 1993, pp.6-389.

[5] Diaz, L.F.; Savage G.M.; Eggerth, L.L.; Rosenberg, L. *Solid Waste Management;* ISBN-92-807-2676-5, UNEP International Environmental Technology Centre,CalRecovery, Inc, USA and Japan, 2005, Vol. 1, pp. 1-6.

[6] Lavee, D.; Regey, U.; Zemel, A. *J. Environ Manage.* 2009, *vol 90*, 3599-3606.

[7] Hamoda, M.F.; Abu Qdais, H.A.; Newham, J. *Resour Conserv Recy.* 1998, *vol 23,* 209-223.

[8] Hassan, H.F. Construction and Building Materials. 2005, *vol 19,* 91-98.

[9] Duan, H.; Huang, Q.; Wang, Q.; Zhou B.; Li, J. *J. Hazard. Mater.* 2008, *vol 158,* 221-227.

[10] LaGrega, M.D.; Buckingham, P.L.; Evans, J.C. *Hazardous Waste Management,* McGraw-Hill International Editions, Civil Engineering Series, 1994, pp. 1-23.

[11] Hazra, T.; Goel, S. *Waste Manage.* 2009, *vol 29,*470-478.

[12] Al-Khatib, I.A.; Monou, M.; Abu-Zahra, A.S.F.; Shaheen, H.Q.; Kassinos, D. *J Environ Manage.* 2010, *vol 91,* 1131-1138.

[13] Magrinho, A.; Didelet, F.; Semiao, V. *Waste Manage.* 2006, *vol 26,* 1477-1489.

[14] Minghua, Z.; Xiumin, F.; Rovetta, A.; Qichang, H.; Vicentini, F.; Bingkai, L.; Giusti, A.; Yi, L. *Waste Manage.* 2009, *vol 29,* 1227-1233.

[15] Uriarte, F.A.Jr. *Solid Waste Management:Principles and Practices;* ISBN-978-971-542-557-5, Quezon, Philippines, 2008, Vol. 1, pp. 1-61.

[16] Kofoworola, O.F. *Waste Management.* 2007, *vol 27,* 1139-1143.

[17] Bagchi, A. *Design of Landfills and Integrated Solid Waste Management*; ISBN-0-471-25499-1, John Wiley&Sons, Inc, New Jersey, USA, 2004, Vol. 1, pp. 1-39.

[18] Bari, Q.H.; Koenig, A. *Resour Conserv Recy.* 2001, *vol 33,* 93-111.

[19] Daskalopoulos, E.; Badr, O.; Probert, S.D. *Resour Conserv Recy.* 1998, *vol 24,* 33-50.

[20] Louis, G.; Shih, J.S. *Socio-Economic Planning Sciences.* 2007, *vol 41*, 61-89.

[21] Thomas Hope, E. *Solid Waste Management: Critical Issues for Developing Countries;* ISBN-976-8125-43-8, Canada, 1998, Vol. 1, pp. 1-8.

[22] Clift, R.; Doig, A.; Finnveden, G. *Trans IChemE.* 2000, *vol 78*, 279-287.
[23] Williams, P.T. *Waste Treatment and Disposa;* ISBN-0-470-84912-6 (HB) John Wiley&Sons, England, 2005, Vol. 1, pp. 1-15.
[24] Phillips, P.S.; Gronow, B.; Read, A.D. *Resour Conserv Recy.* 1998, *vol 23,* 127-161.
[25] Glavic, P.; Petek, J. *Resour Conserv Recy.* 1996, *vol 17,* 169-188.
[26] Buenrostro, O.; Bocco, G. *Resour Conserv Recy.* 2003, *vol 39,* 251-263.
[27] Alam, R.; Chowdhury, M.A.I.; Hasan, G.M.J.; Karanjit, B.; Shrestha, L.R. *Waste Manage.* 2008, *vol 28,* 1088-1097
[28] Tchobanoglous G.; Krwith, F. *Handbook of Solid Waste Management*; ISBN-0-07-135623-1, McGraw-Hill, USA, 2002, Vol. 1 pp. 1-11.
[29] El-Haggar, S. *Sustainable Industrial Design and Waste Managemen;* ISBN-13 978-0-12-373623-9, Elsevier Inc, USA, 2007, Vol. 1 pp. 1-6.
[30] Agarwal, A.; Singhmar, A.; Kulshrestha, M.; Mittal, A.K. *Resour Conserv Recy.* 2005, *vol 44,* 73-90.
[31] Adams, S.K.; Even, J.C.Jr; Blewett, T.J. *Resour Conserv Recy.* 1996, *vol 17,* 153-167.
[32] Hargreaves, J.C.; Adl, M.S.; Warman, P.R. *Agr Ecosyst Environ.* 2008, *vol 123,* 1-14.
[33] Xi, B.; Zhang, G.; Liu, H. *J. Hazard. Mater.* 2005, *vol B124,* 165-172.
[34] Elango, D.; Thinakaran, N.; Panneerselvam, P.; Sivanesan, S. *Applied Energy.* 2009, *vol 86,* 663-668.
[35] Slater, R.A.; Frederickson, J. *Resour Conserv Recy.* 2001, *vol 32,* 359-374.
[36] Cheng, H.; Hu, Y. *Environ Pollut.* 2010, *vol 158,* 2809-2814.
[37] Veli, S.; Öztürk, T.; Dimoglo, A. *Separation and Purification Technology.* 2008, *vol 61*, 82-88.
[38] Slack, R.J.; Gronow, J.R.; Hall, D.H.; Voulvoulis N. *Environ Pollut.* 2007, *vol 146,* 501-509.
[39] Erses, A.S.; Onay, T.T.; Yenigun, O. *Bioresource Technol.* 2008, *vol 99,* 5418-5426.
[40] Read, A.D.; Phillips, P.; Robinson, G. *Resour Conserv Recy.* 1997, *vol 20,* 183-205.

In: Municipal Solid Waste
Editors: G. Konno and C. Machado

ISBN 978-1-62100-257-4

Chapter 4

NO AND N_2O EMISSIONS DURING FLUIDISED BED COMBUSTION OF LIGNITE AND MUNICIPAL SOLID WASTE (MSW)

Suthum Patumsawad*
Department of Mechanical Engineering
King Mongkut's University of Technology North Bangkok,
Bangsue Bangkok, Thailand

ABSTRACT

An experimental investigation on the co-firing characteristics of Thai lignite and municipal solid waste (MSW), with emphasis on the emissions of major nitrogenous-species, has been carried out in a bubbling fluidised bed. The lignite used was of 3% sulphur and the MSW was simulated using materials representing the main compositions of high moisture MSW, that is, a mixture of vegetable, paper, bark, plastics and crushed brick. The effects of co-combustion on temperature distributions, combustion zone and carbon combustion efficiency were investigated, and their implications on the influence of the mass fraction of MSW in the MSW/lignite mixture on the emissions of NO, N_2O, NH_3 and HCN, measured using an FTIR gas analyzer, were discussed. Tests were run with MSW mass fraction varied between 0 and 40% and excess air between 40 and 100%. The results show that the addition of MSW results in a drop in bed temperature and a slight decrease in carbon combustion efficiency, due primarily to the high moisture content of MSW. The high volatile content of the MSW promotes combustion in the freeboard, leaving less fuel particles for in-bed combustion. This phenomenon is most obvious in the case of 20% MSW co-combustion, where the synergistic effect of high volatility on the reactivity of the mixture is believed to be the highest. The emission of NO generally increases with increasing mass fraction of MSW due mainly to the increase in the release of NH_3 from the MSW and a decrease in catalytic char particles for NO reduction. NO appears to be minimised at 20% MSW. N_2O decreases slightly to a minimum at 20% MSW due to temperature decrease, but increases as excess air increases further because of increasing presence of CH_4 and NO. The emission of NH_3 is much

* Tel. +66 2913 2500 ext 8322 Fax. +66 2586 9541; E-mail: stt@kmutnb.ac.th.

more significant than HCN. Overall, the total fuel-nitrogen converted as NO, N_2O, HCN and NH_3 increases with MSW mass fraction, except at 20%.

In addition, the comparison of 35% waste moisture and 60% waste moisture were preformed. The results show that the bed temperature in the case of 35%-moisture is higher than for in case of 60% due to the difference of physical properties of the fuel. The combustion efficiency for the case of 35% waste moisture is higher than that for 60% moisture due to higher bed temperature at lower waste moisture. The synergistic effect of the co-combustion of lignite with the drier MSW reduces the emission of CO leading to increase in combustion efficiency. The fuel mixture with low-moisture in waste gives higher fuel-N conversion to NO whereas the fuel-N conversion to N_2O is higher for higher moisture waste, particularly at high excess air.

Keywords: Fluidised Bed, Co-combustion, Lignite, MSW, Nitrogenous Species Emissions

1. INTRODUCTION

The disposal of municipal solid waste (MSW) is a growing problem worldwide because of the ever-increasing volume of the waste and the potential negative impact it has on the environment, health and safety. Furthermore, because of growing public resistance, even currently accepted as proper MSW disposal methods and practices can be highly controversial and in many cases very expensive. Therefore the adoption of any MSW disposal option requires good public acceptance and careful consideration of the cost.

Among several MSW disposal options used to date, incineration is becoming increasingly more attractive since it provides maximum volume reduction and energy recovery. MSW can be burnt relatively efficiently with energy recovery in several types of incinerators, particularly fluidised bed combustors, which are becoming more popular because of its ability to handle wastes of widely varied properties and the many advantages in controlling emission (Wheeler et al., 1995, McGowin and Howe, 1994). Recent experiences, commercial and experimental, indicate that MSW can be co-fired with coal or other conventional fuels, offering higher energy conversion efficiency, lower capital investment, and lower risk than the building of dedicated incinerators to recover energy (Manninen et al., 1997, Frankenhaeuser et al., 1994). This is particularly so with high moisture and low calorific value MSW. In addition, co-firing MSW with a high sulphur coal typically reduces sulphur dioxide (SO_2) emission relative to those firing 100 percent coal (McGowin and Hughes, 1992, Desroches-Ducarne et al, 1998). In Thailand, where there is a severe lack of proper disposal for its estimated 45,000-ton per day high moisture MSW and where SO_2 emission resulting from the burning of high-sulphur lignite is of serious concern, co-firing is believed to be a suitable strategy. However, depending on the waste composition and operation conditions, co-firing may either increase or decrease the emission of other pollutants such as the nitrogenous gases. Since the major nitrogenous pollutant gases are NO and N_2O which, respectively, causes acid rain and contributes to the greenhouse effect and ozone layer depletion, it is desirable that investigations be carried out on the subject.

The aim of this paper is to present the results obtained from an experimental study on the affect of co-firing of high-sulphur Thai lignite and high-moisture MSW in a fluidised bed combustor to the emissions of nitrogen containing gas species: NO and N_2O. The characteristics of co-firing, including temperature distribution, combustion regimes, and carbon combustion efficiency are described and the affect of MSW mass fraction to the emission of nitrogeneous species are illustrated.

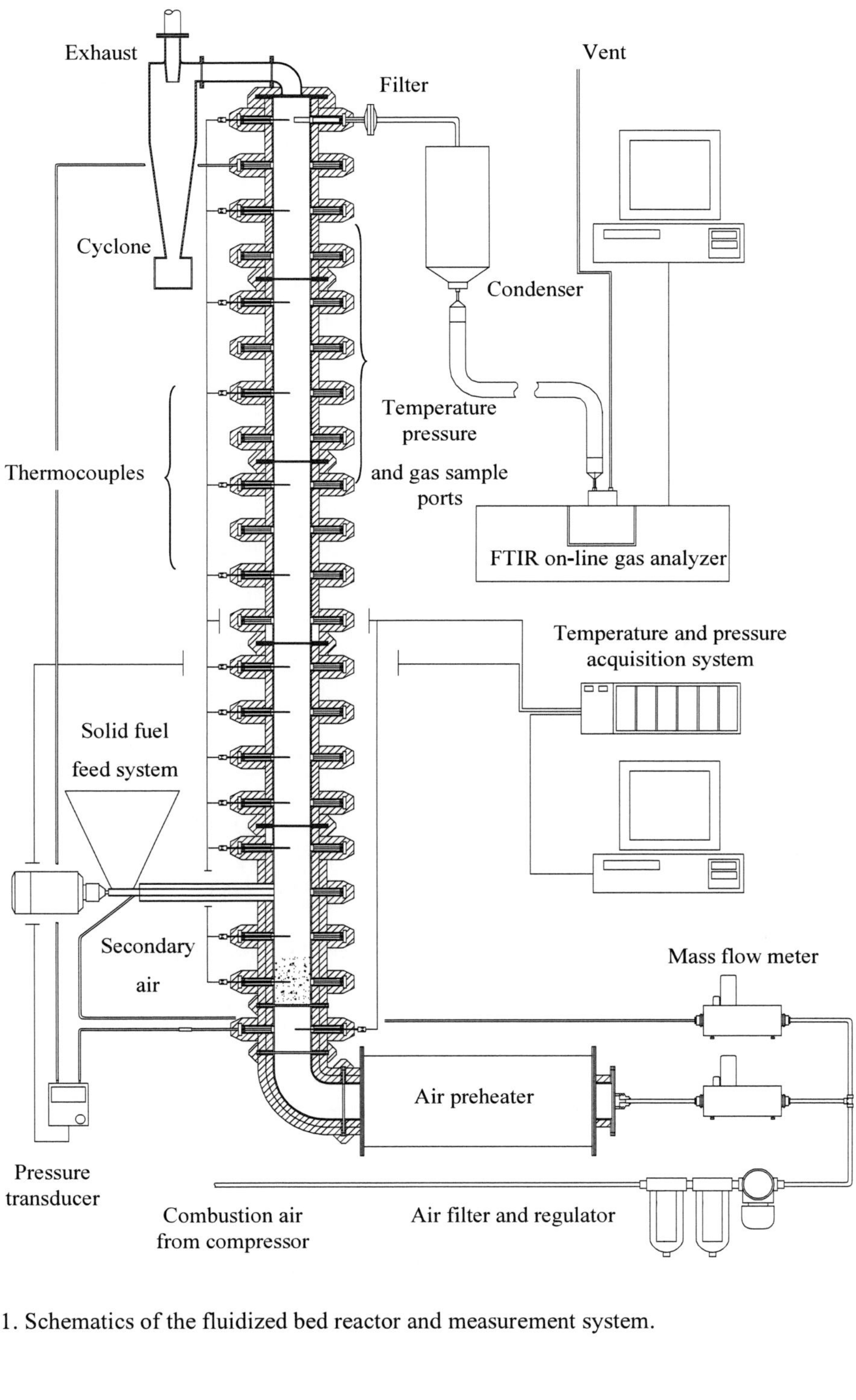

Figure 1. Schematics of the fluidized bed reactor and measurement system.

2. Experimental Set-up

The experiments were performed in a bubbling fluidised bed reactor shown schematically in Figure 1. The reactor is a stainless-steel cylindrical tube of 10.5-cm ID and 3.0 m in height, with outside ceramic fibre insulation. Silica sand, 2,600 kg/m^3 density and 200μm mean diameter, was used as the bed material. Fluidising air from a compressor was introduced into the bed through an orifice-type distributor. The fuel mixture was fed on the bed at 37.5 cm above the air distributor plate and the feed rate was such that a nominal thermal input of 3.3 kW – equivalent to 0.75 kg/h of lignite -- was maintained for all experiments. The fluid bed was heated by hot air, which was preheated by a 6-kW electric heater. The temperature of the bed was typically 400°C. This temperature is sufficient for ignition of lignite. The operating bed temperature was in the range of 800-920°C. Entrained particles were separated from the exhaust gas by a cyclone. Gas samples could be taken through sampling probes at various positions along the bed. In this experiment, the gas samples were taken continuously at 262.5 cm above the air distributor, just before the freeboard exit. All the samples were taken in the axis of the reactor. The sampled gas was pumped through a filter and condenser before entry into an on-line Fourier Transform Infrared gas analyzer (FTIR) for continuous measurement of NO, N_2O, HCN, NH_3, CO, CO_2, and CH_4.

In order to maintain a consistent set of fuel properties, the MSW used was simulated by using materials that may substitute the main compositions of collected waste. In these experiments the MSW was composed of a mixture of vegetable, paper, bark, plastics and brick. The size of the materials ranged of 1-5 mm. The lignite, contributed by the Electricity Generating Authority of Thailand, was crushed and sieved to 1-5 mm. Tables 1 and 2 show the proximate and ultimate analyses of lignite and MSW, respectively.

Results and Discussion

The mass fraction of MSW in the MSW/lignite mixture was varied from 0 to 40% at 10% intervals. For each MSW mass fraction, the excess air was varied from 40 to 100% at 20% intervals. Table 3 shows the experimental conditions used in all tests.

Table 1. Proximate analysis (%wt) of lignite and MSW

	Moisture	VM (dry)	Fixed carbon (dry)	Ash (dry)	LHV (MJ/kg)
Lignite	16	35 (42)	37 (44)	12 (14)	15.7
Model MSW	60	26 (65)	4 (10)	10 (25)	5.3

Table 2. Ultimate analysis of lignite and MSW (%wt) daf

	C	H	O	N	S	Ca
Lignite	49.8	4.1	40.8	1.68	2.96	0.76
Model MSW	47.81	6.73	43.18	1.6	0.06	0.57

Table 3. Experimental conditions and measured concentrations of nitrogenous species (converted to 6% O_2)

ni WSM% erutxiM	etar deef leuF (h/gk)	riA ssecxE (%)	laicifrepuS yticoleV saG s/m()	(mpp) snoitartnecnoC saG			
				ON	N_2O	NCH	NH_3
0	0.750	40	0.41	122	37	28	145
	0.750	60	0.47	171	70	27	53
	0.750	80	0.53	284	75	26	39
	0.750	100	0.59	439	75	23	24
10	0.800	40	0.45	139	32	25	285
	0.800	60	0.51	175	56	25	217
	0.800	80	0.57	313	65	24	78
	0.800	100	0.64	365	72	19	28
20	0.860	40	0.48	107	20	24	264
	0.860	60	0.55	138	39	24	145
	0.860	80	0.62	177	47	23	60
	0.860	100	0.69	213	49	19	24
30	0.935	40	0.53	191	29	26	371
	0.935	60	0.61	243	42	26	309
	0.935	80	0.68	367	55	25	108
	0.935	100	0.76	546	66	22	78
40	1.020	40	0.58	225	34	22	467
	1.020	60	0.67	339	43	21	383
	1.020	80	0.75	541	57	20	138
	1.020	100	0.83	576	62	19	114

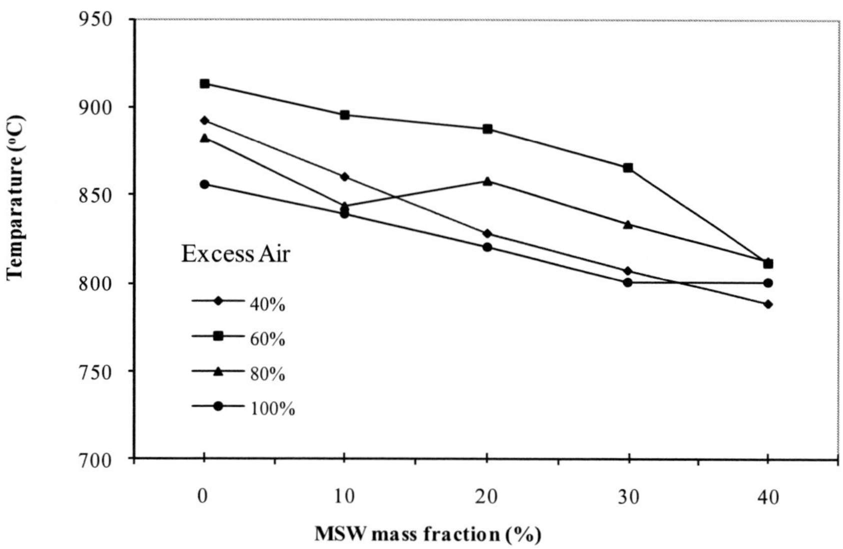

Figure 2. Effect of MSW mass fraction on bed temperature.

3.1. Effects on Temperature and Combustion Efficiency

The effect of MSW mass fraction on the mean bed temperature, measured at 7.5 cm above the air distributor plate in a bed of approximately 30 cm in height when fluidised, is shown in Figure 2. Despite constant thermal input (at 3.3 kW), the bed temperature drops almost linearly, at constant excess air, with increasing portion of MSW in the mixture. This is due to the high moisture and high volatile content of the waste, which leaves smaller amount of char in the bed. The axial temperature distribution of the fluidised bed for different MSW mass fractions at a constant excess air of 60% is shown in Figure 3. It is interesting to note that, unlike other cases, the peak temperature for the case of 20% MSW occurs at the so-called "splash zone", which is the fuel injection zone (located at 37.5 cm above the air distributor plate), rather than in the bed. A plot of the temperature difference between the splash zone and the bed ($T_{splash} - T_{bed}$), as shown in Figure 4, further reinforces this distinctive feature, where it is seen that the temperature difference at 20% MSW is the smallest and is positive for all excess air levels investigated. This implies that, at this MSW mass fraction, active combustion takes place in the vicinity of splash zone or disengagement zone. The fact that combustion in the splash zone is more likely in the case of 20% MSW than any other case is confirmed by the result of a thermogravimetric (TGA) experiment shown in Figure 5. The differential thermogravimetric (DTG) curves for different mixtures, ranging 100% lignite to 100% MSW, clearly illustrate that at 20% MSW, devolatilisation occurs at the lowest temperature and the rate of mass loss is relatively higher. This suggests that the reactivity of the mixture is enhanced by the presence of MSW due to its higher volatile matter content -- a synergistic effect of co-firing two different materials as reported by Sorum *et al.* (2001). In an investigation on the interactions between different paper and plastic components in a mixture, they found that the reactivity of the cellulosic matter increased with the presence of plastics. Because of this, it may be concluded that, due to high reactivity at 20% MSW, rapid devolatilisation occurs in the vicinity of the splash zone before the fuel particles reach the bed. This salient behaviour is to have an important bearing on the emission of nitrogen containing species, as will be described in the following section.

It should also be noted in Figure 3 that the temperature in the bed for 100% lignite combustion is higher than that for MSW/lignite mixture, while the reverse is true in the upper part of the freeboard. This implies that, because of the higher volatile and lower fixed carbon content in MSW (Table 1), less fuel particles are burnt in the bed in the case of co-combustion. This phenomenon is in agreement with that of Cliffe and Patumsawad (2001), who investigated the co-firing of olive oil, which contains high volatile matter with coal.

The variations of carbon combustion efficiency and emissions of CH_4, and CO are shown in Figs 6-8 as a function of MSW mass fraction. Since the emission of CH_4 can be substantial at low excess air, we take into account the carbon emitted as CH_4 in estimating the carbon combustion efficiency, as follow:

$$\eta_{CE} = \frac{\%CO_2}{\%CO_2 + \%CO + \%CH_4}$$

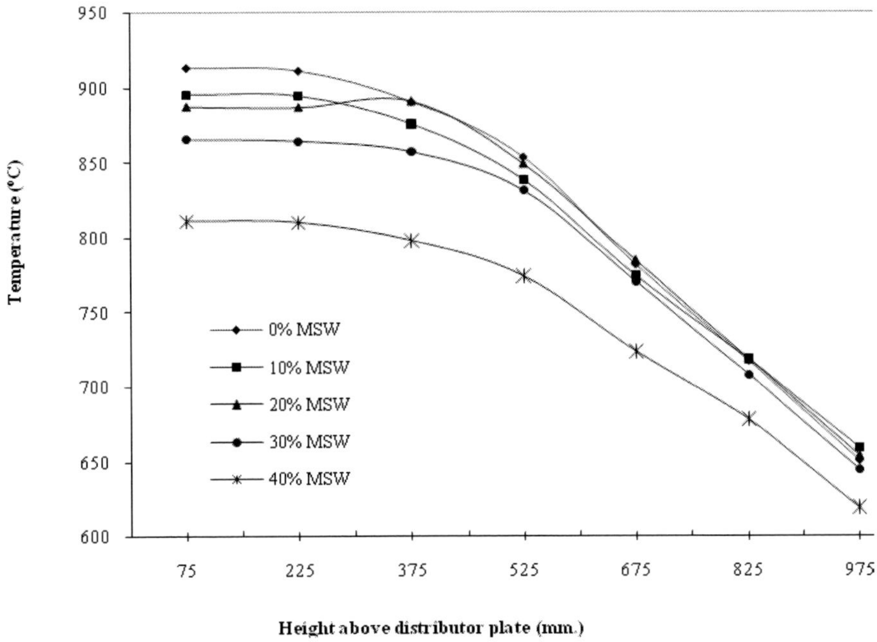

Figure 3. Effect of MSW mass fraction on axial temperature distribution (at 60% excess air).

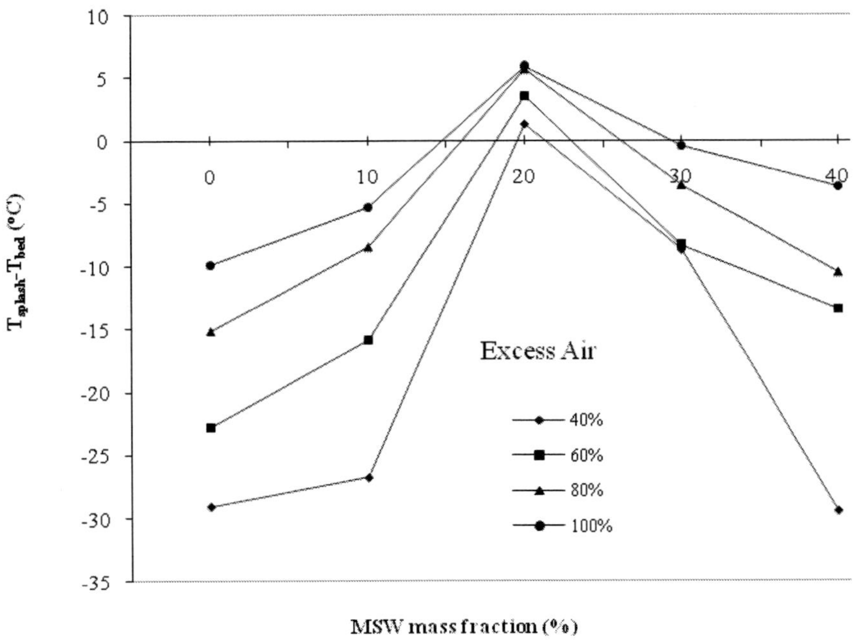

Figure 4. Effect of MSW mass fraction on T_{splash} - T_{bed}.

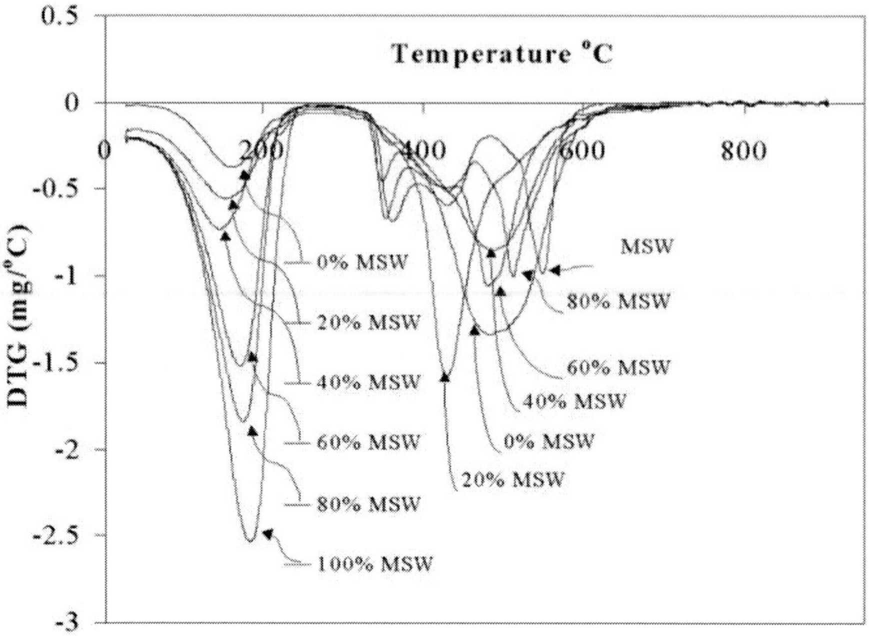

Figure 5. DTG curves for varying mass fraction of MSW in MSW/lignite mixture.

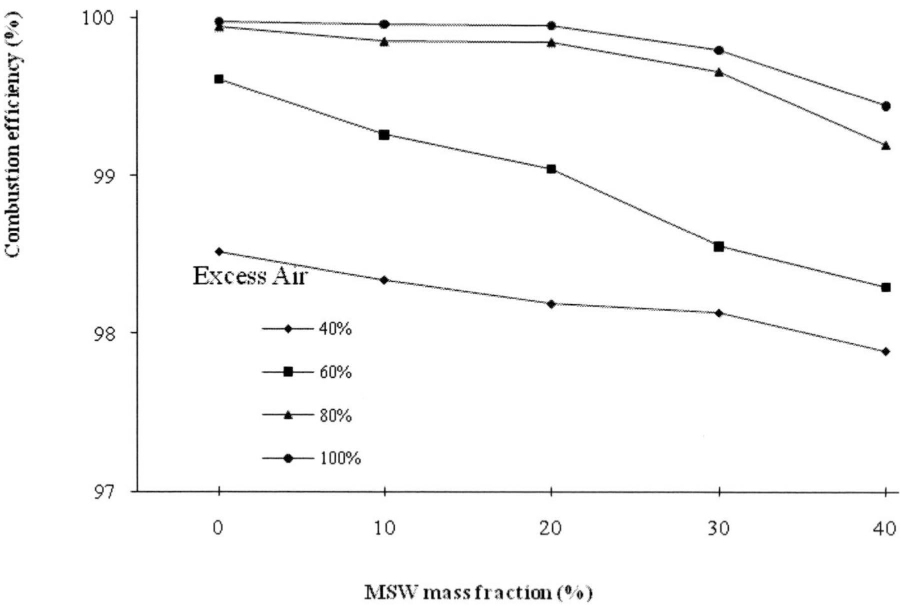

Figure 6. Effect of MSW mass fraction on carbon combustion efficiency.

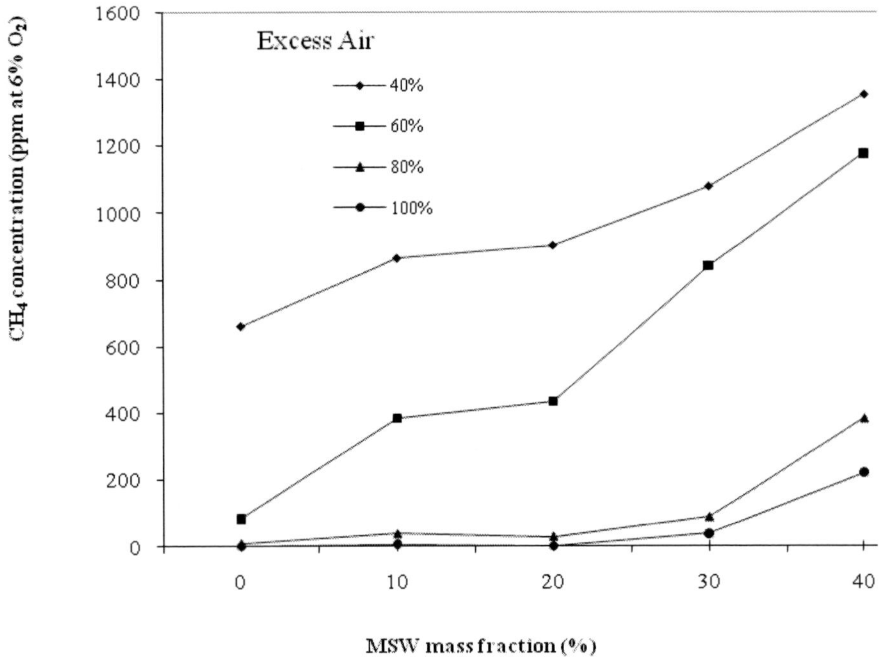

Figure 7. Effect of MSW mass fraction on emission of CH_4.

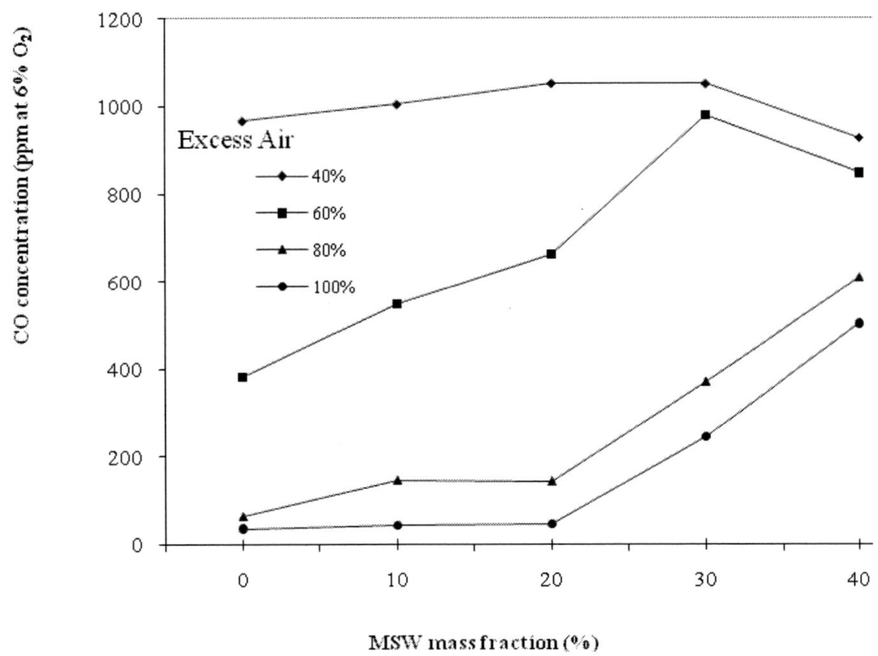

Figure 8. Effect of MSW mass fraction on emission of CO.

It can be seen in Figure 6 that the carbon combustion efficiency decreases with the amount of MSW added, particularly when the fraction of MSW is at 30% and above. The decrease can be directly attributed to the drop in bed temperature resulting from increased moisture of the fuel mixture as discussed previously. Moisture can also inhibit hydrocarbon oxidation (Desroches-Ducarne *et al.,* 1998 and Bromly *et al.*, 1988), resulting in increased emissions of CH_4 and CO as shown in Figures 7 and 8. However, the effects of MSW addition on combustion efficiency, CO and CH_4 are relatively insignificant at low fractions of MSW, particularly at high excess air. This is probably due to the rapid combustion of volatiles, which is sufficient to compensate for the drop in bed temperature caused by increased moisture. The maximum reduction in efficiency is 1.3%.

The Emission of NO

Under fluidised bed conditions, NO and N_2O emissions originate from the nitrogen released with the volatiles and from nitrogen in the char, so that the formation of thermal NO is considered to be insignificant (Furusawa et al, 1978). Therefore in this paper, the emissions of nitrogenous species are presented in terms of the percentage of fuel-nitrogen (fuel-N) converted.

The emissions of NO and N_2O are shown in Figures 9 and 10 respectively. Since it is well known that NO is formed from both HCN and NH_3, while N_2O is formed exclusively from HCN (Loffler et al, 2001), the emissions of NH_3 and HCN are also plotted in Figures 11 and 12, respectively. For the range of conditions investigated, NO emission ranges from 107 to 576 ppm (@ 6% O_2), representing 3.7 to 20.4 % of fuel-N converted. In general NO increases with increasing fraction of MSW, except for the case of 20% MSW where a distinctive drop is seen for all percentages of excess air investigated, particularly at high excess air. The increase in NO, despite a drop in bed temperature shown in Figure 2, could be due to a number of reasons.

Firstly, as the fraction of MSW increases, there may be more NH_3 released for conversion to NO. This is evidenced by the fact that the emission of NH_3 increases with increasing fraction of MSW (Figure 11). Desroches-Ducarne, *et al.* (1997) explained that nitrogen contained in wood (~ 60% of the combustibles in MSW) exists in the form of simple protein, which is known to be released as NH_3 when burned. With increasing concentrations of CO and CH_4 (Figures 6 and 7) as MSW increases, more O and OH radicals are produced, which serve to promote NO formation through NH_3 oxidation (Wargadalam et al. 2000).

Secondly, as discussed earlier, the combustion of high volatile content MSW leaves less char particles in the bed that would otherwise serve to reduce NO through the following mechanisms:

$$NO + C \longrightarrow \tfrac{1}{2} N_2 + CO$$

$$NO + CO \xrightarrow{\text{Char}} \tfrac{1}{2} N_2 + CO_2$$

Thus, although CO is seen to increase with MSW fraction (Figure 8), the lesser char particles available as catalyst inhibit the role of CO as an agent for the reduction of NO (Philippek and Werther, 1997). An estimation of the fraction of fuel-N converted to N_2 (total fuel-N minus the sum of N in NO, N_2O, HCN and NH_3) as shown in Figure 13 for the case of 60% excess air confirms that, relative to NO, the conversion to N_2 is less favoured as MSW increases.

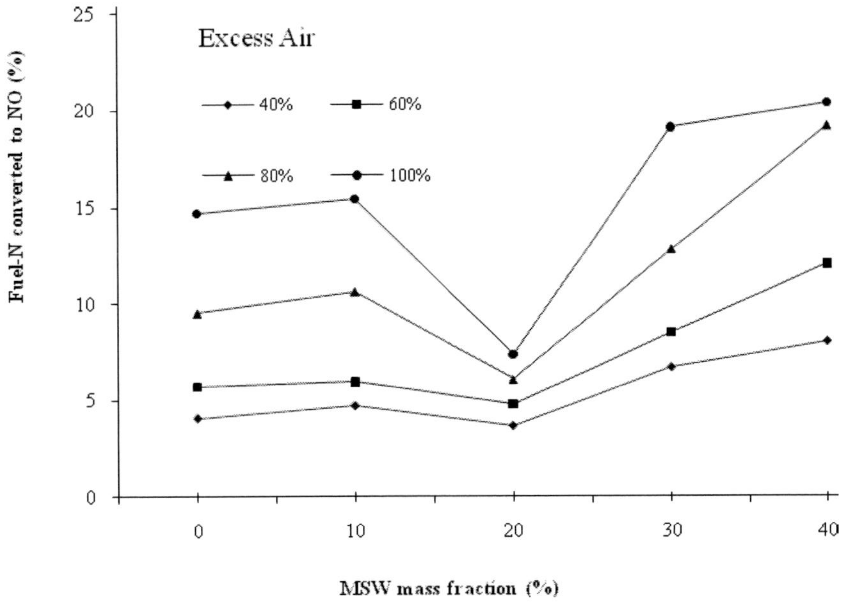

Figure 9. Effect of MSW mass fraction on emission of NO.

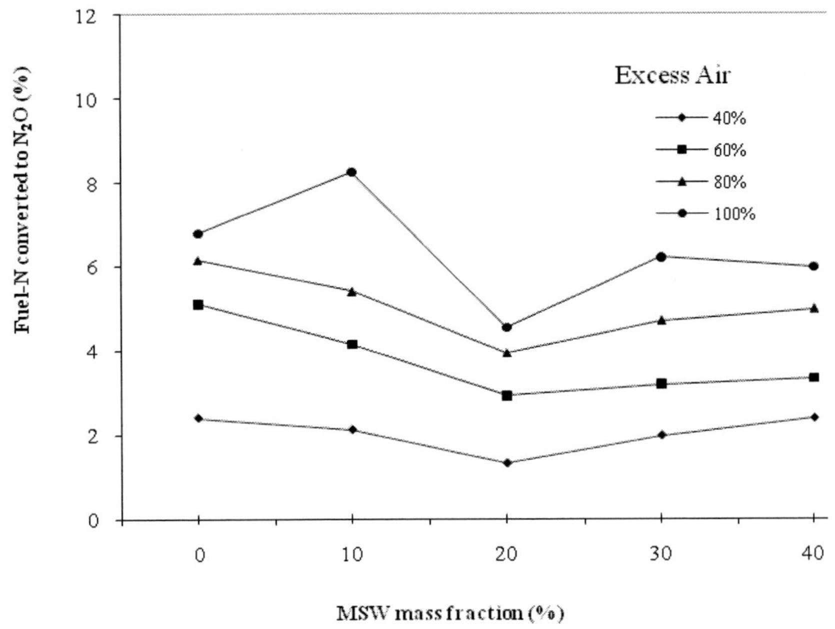

Figure 10. Effect of MSW mass fraction on emission of N_2O.

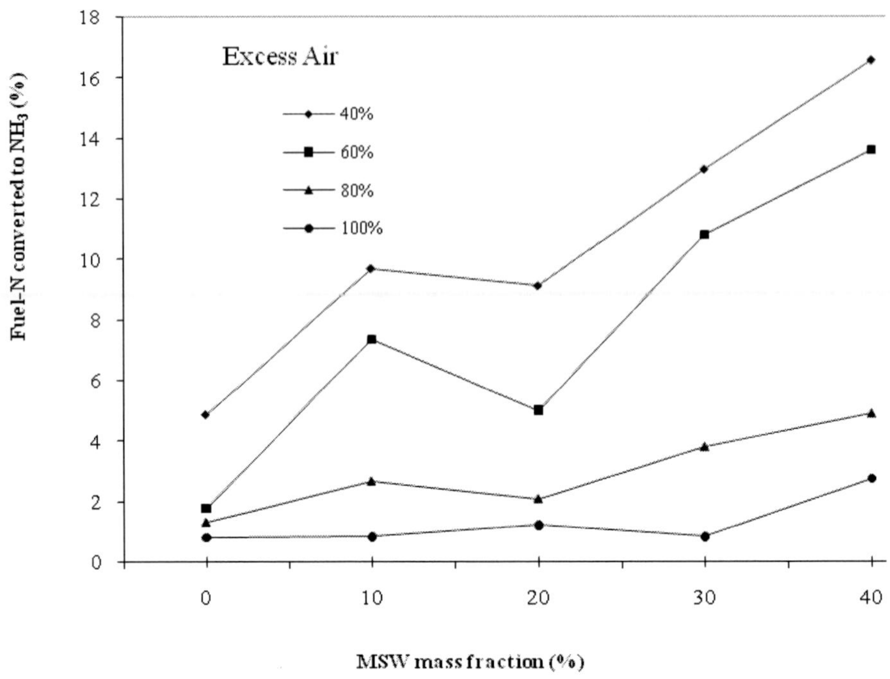

Figure 11. Effect of MSW mass fraction on emission of NH_3.

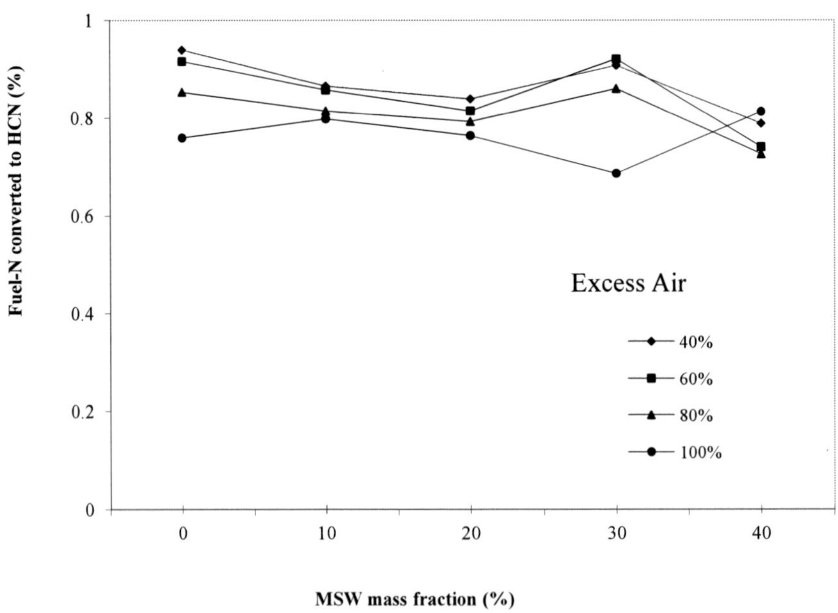

Figure 12. Effect of MSW mass fraction on emission of HCN.

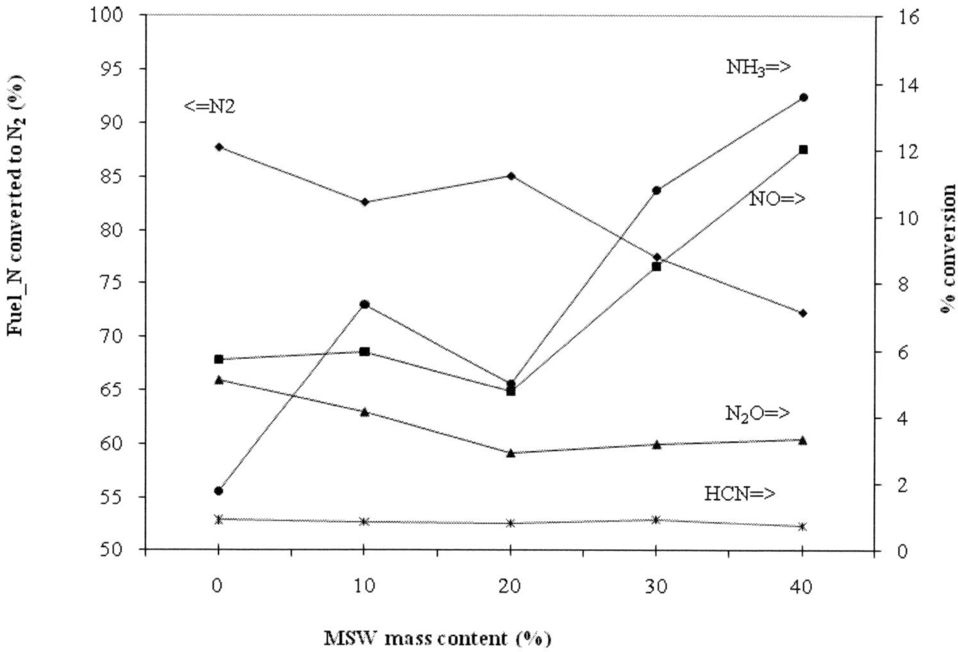

Figure 13. Emission of nitrogenous species at 60% MSW. (N_2 calculated as Fuel_N-($NO+N_2O+NH_3+HCN$)).

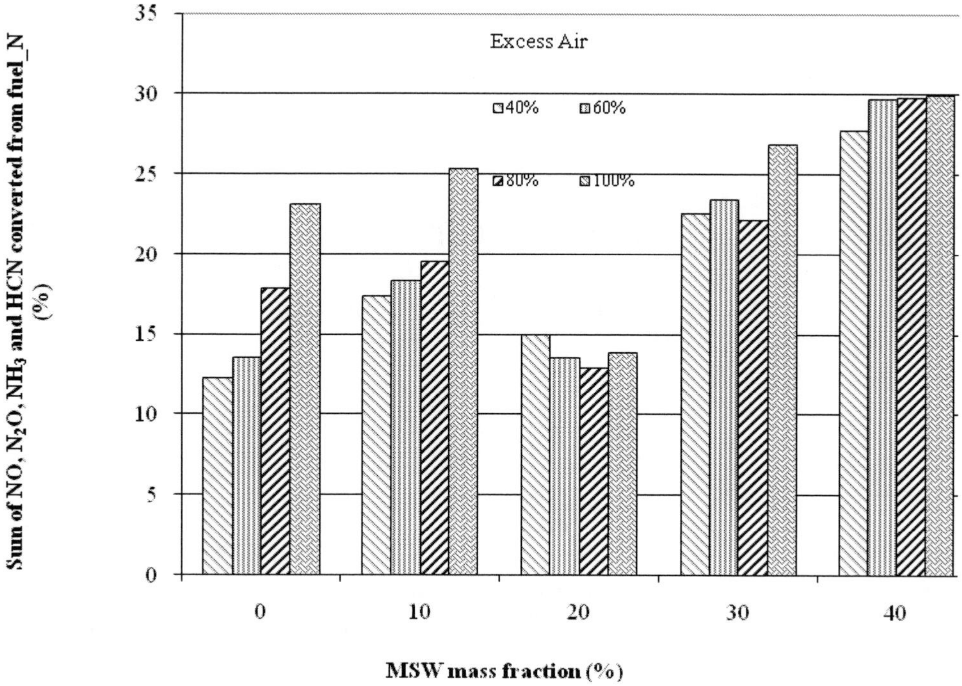

Figure 14. Total conversion of fuel-N as a function of MSW fraction and excess air.

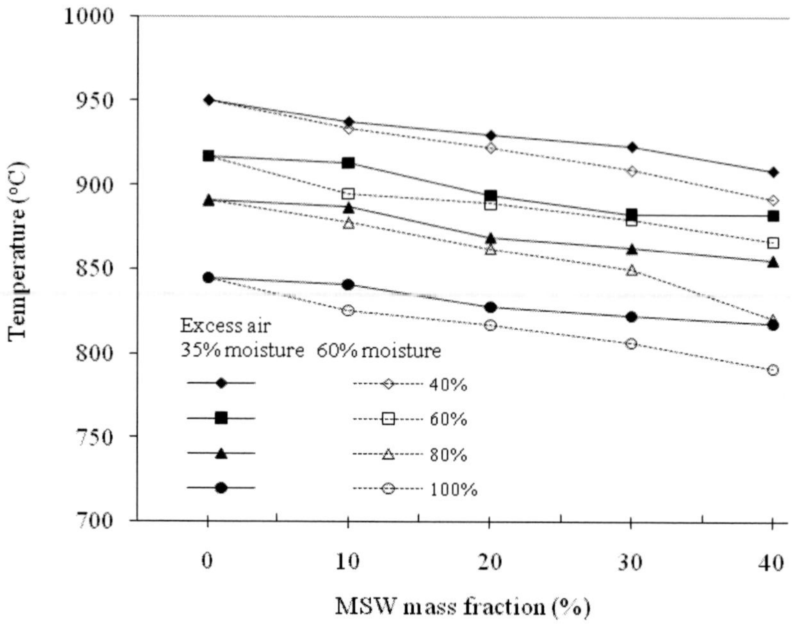

Figure 15. Effect of moisture and excess air on bed temperature as a function of MSW mass fraction.

Thirdly, the increase of MSW contributes more moisture to the fuel mixture, leading to more H and OH radicals, which tend to accelerate the production of NO by increasing the rate of production of N atoms (Bowman, 1975).

The drop in NO at 20% MSW is related to the optimal reactivity of this particular fuel mixture as explained in the previous section, which in turn affects the selectivity of NH_3 conversion to either NO or N_2. As typified in Figure 13, the drop in NO corresponds to a drop in NH_3 but a rise in N_2. The reason is that, because more combustion takes place in the vicinity of the splash zone as discussed previously, there is lower concentration of fuel particles in the said zone to catalyse the NH_3-to-NO oxidation. The increased reaction rate may also have enhanced NO reduction by NH_3 (Hampartsoumian and Gibbs, 1982, and Jahkola *et al.*, 1991).

Although NO is also formed from HCN, the fact that the emission of the latter remains relatively constant when MSW is added (Figure 12) suggests that the variation of NO is due primarily to NH_3 conversion.

3.3. The Emission of N_2O

Figure 10 shows the variation of N_2O emission as a function of MSW mass fraction at constant excess air. In general, N_2O decreases to a minimum at 20% MSW and then increases slightly as the fraction of MSW increases. The reasons for this phenomenon can be given as follow:

It has been shown that N_2O formation from HCN increases with temperature (Wargadalam et al. 2000). Thus N_2O decreases with the decrease in temperature as MSW increases, until the MSW fraction reaches 20%. The decrease may also be caused by the homogeneous destruction of N_2O by the increase of H and OH radicals (Leckner and Gustavsson, 1991) due to increased moisture. As the fraction of MSW increases further, however, the role of carbon-containing species such as CH_4 appears to have played an important role in N_2O formation. It is well known that N_2O formation at temperatures above 800 °C strongly depends on the concentration of O and OH radicals, as well as the intermediate gas species NCO, all of which can be increased during the combustion of CH_4 (Wartha *et al.* 2000). Therefore the formation of N_2O may be enhanced with increasing MSW as CH_4 is also seen to increase beyond 20% MSW (Figure 7). The higher conversion to N_2O can also be a consequence of higher instantaneous NO concentration as the latter may further react with bound nitrogen to produce N_2O according to the following heterogeneous reaction (Chen *et al.* 2001):

$$\text{Char\_N} + NO \longrightarrow N_2O + C$$

It should also be noted that the emission of N_2O is much less significant than NO under the same excess air (see Figure 13 for example).

Overall, the total fuel-N converted to NO, N_2O, HCN and NH_3 increases with the addition of MSW, except at 20% MSW (Figure 14). This is because, for a high-volatile fuel such as MSW, the conversion of fuel nitrogen to HCN and NH_3 - the two precursors of NO, N_2O and N_2 - during devolatilisation is expected to be more significant than conversion from char particles.

3.4. The Effect of Moisture in MSW

To study the effects of moisture in MSW on combustion characteristics and flue gas emissions, experiments on 35%-moisture MSW was performed and the results were compared with the combustion of 60%-moisture.

3.4.1. Temperature Distribution and Combustion Efficiency

Figure 15 shows the effect of waste moisture on bed temperature. It can be seen that for 35%-moisture MSW, the temperature drop resulting from increased MSW mass fraction in the fuel mixture is smaller, averaging 8°C for every 10% increase in waste mass fraction, as compared to 15°C for the case of 60% moisture. The higher bed temperature in the case of 35%-moisture is due to the fact that the percentage of fixed carbon is higher (10.6% against 4.0%) and that less heat is required to evaporate the moisture. The temperature difference for the two cases is even more pronounced at higher waste mass fractions.

Figure 16 shows a typical temperature distribution in the case of 30% MSW and 80% excess air. Inside the bed, the temperature for the case of 35%-moisture MSW is seen to be higher than the case of 60%-moisture, which in the freeboard, the temperature for case of 60%-moisture is higher. The reasons for this behaviour are similar to that give in Suksankraisorn *et al.* (2002) for 100% lignite combustion and co-combustion of 60%-moisture waste and lignite. That is, with higher moisture in waste the rapid vaporisation of moisture reduces the weight of the waste particles, so that char combustion may begin in the splash zone before they reach the fluid bed, hence higher temperature in the freeboard. Also, as a consequence, less char particles would be left for in-bed combustion, hence lower temperature in the bed.

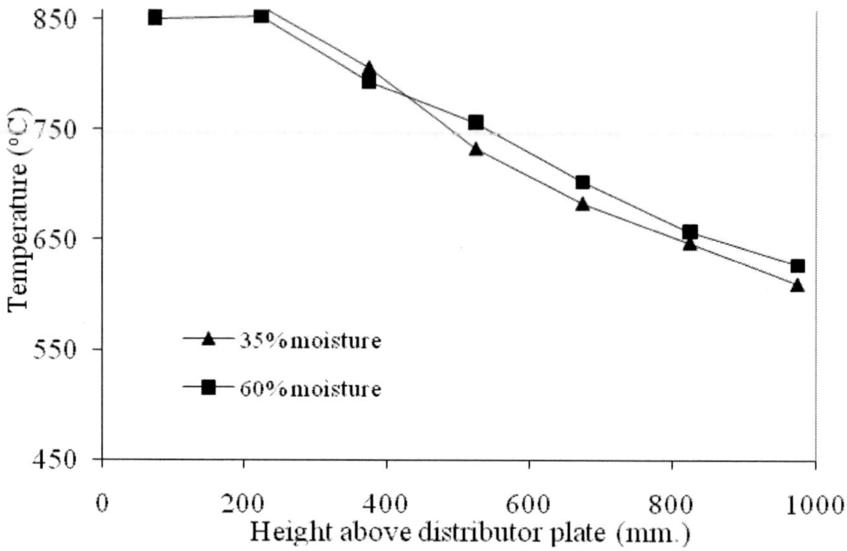

Figure 16. The effect of MSW moisture on axial temperature distribution for the case of 30% MSW and 80% excess air.

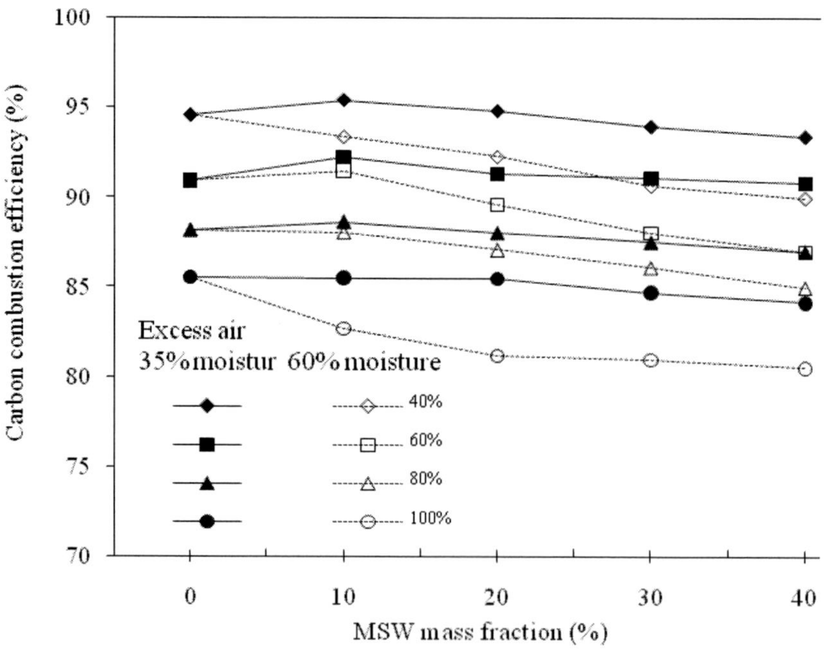

Figure 17. Effect of moisture and excess air on combustion efficiency.

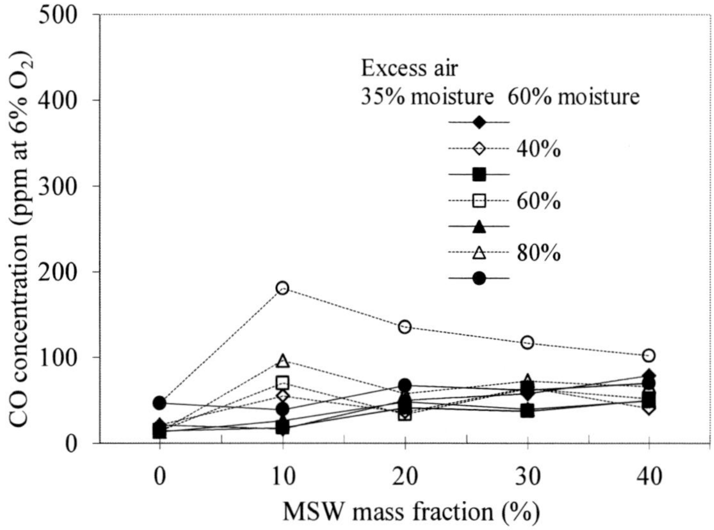

Figure 18. CO emission as a function of MSW mass fraction and excess air.

The effect of waste moisture on carbon combustion efficiency is shown in Figure 17. It is interesting to note that at 35% waste moisture, carbon combustion efficiency increases slightly at 10% MSW mass fraction, and then decreases slightly as the MSW fraction increases further. The occurrence of the peak efficiency at 10% waste corresponds to minimum CO emission at the same waste fraction, as shown in Figure 18. This demonstrates that the synergistic effect of the co-combustion of lignite with the drier MSW occurs at 10% waste fraction rather than at 20% waste fraction.

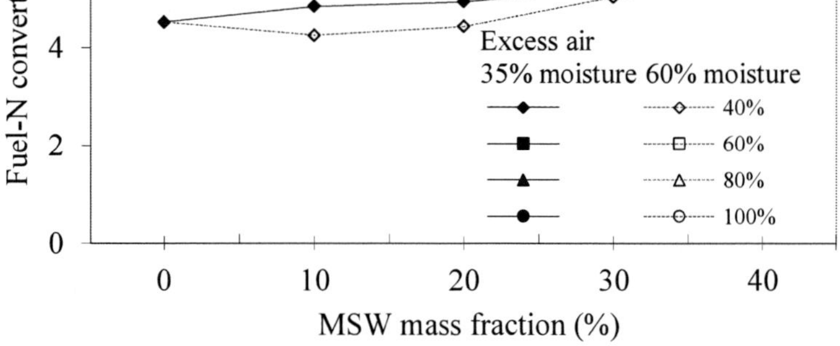

Figure 19. Fuel-N converted to NO as a function of MSW mass fraction.

The combustion efficiency for the case of 35% waste moisture is seen to be higher than that for 60% moisture due to higher bed temperature at lower waste moisture, the efficiency reduction at 40% MSW being 3.4, 3.8, 2.0, and 3.6% for 40, 60, 80, and 100% excess air respectively.

3.4.2. Emission of NO and N_2O

The effect of waste moisture on the percentage of fuel-N conversion to NO is shown in Figure 19. Firstly it is observed that the fuel mixture with 35%-moisture in waste gives higher fuel-N conversion to NO. This is because, with higher moisture, the bed temperature is lower, which may give rise to hydrocarbon free radicals such as CH and CH_2. These radicals will then recycle NO back to CN and HCN, which may in turn be converted to NH_3 by OH and H radicals, and H_2, whose concentrations are higher for high-moisture waste. Furthermore, the catalytic effect of the above hydrocarbon radicals may convert NH_3 and HCN to N_2 rather than NO (Johnsson, 1994).

Secondly, at low MSW fraction (below 20%), the percentage of fuel-N converted to NO for the case of 35% waste moisture is relatively insensitive to changes in MSW mass fraction. This is unlike the case of 60% waste moisture, where there appears to be a minimum fuel-N conversion at MSW fractions around 10-20%. As the MSW fraction increases beyond 20%, NO emission exhibits the same increasing trend for both 35% and 60%-moisture wastes, due to reduced catalytic effect of char particles in the reduction of NO.

Figure 20 shows the effect of waste moisture on the percentage of fuel-N conversion to N_2O. Generally, fuel-N conversion to N_2O is higher for higher moisture waste, particularly at high excess air. This is in contrast with the case of fuel-N to NO conversion. As discussed in previously, the increase in waste moisture leads to the recycling of NO to HCN, which is the precursor of N_2O formation.

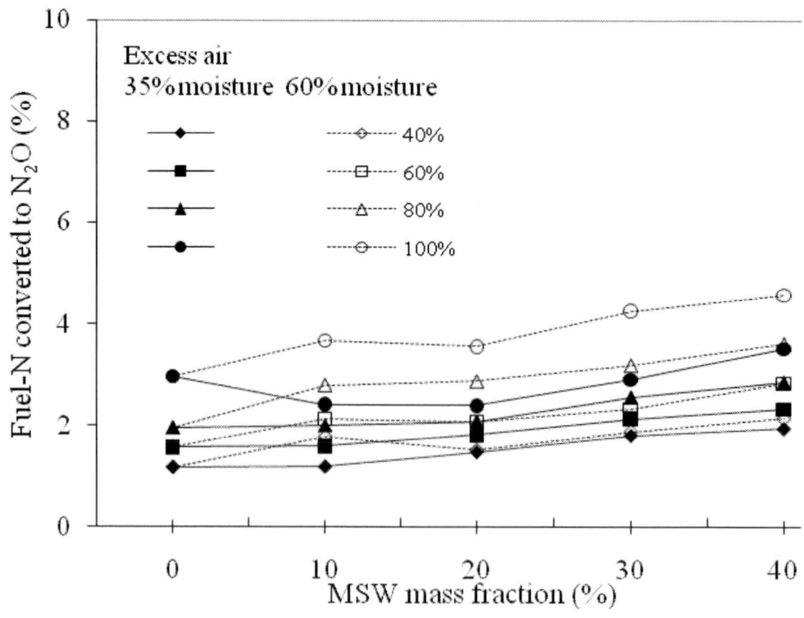

Figure 20. Fuel-N converted to N_2O as a function of MSW mass fraction.

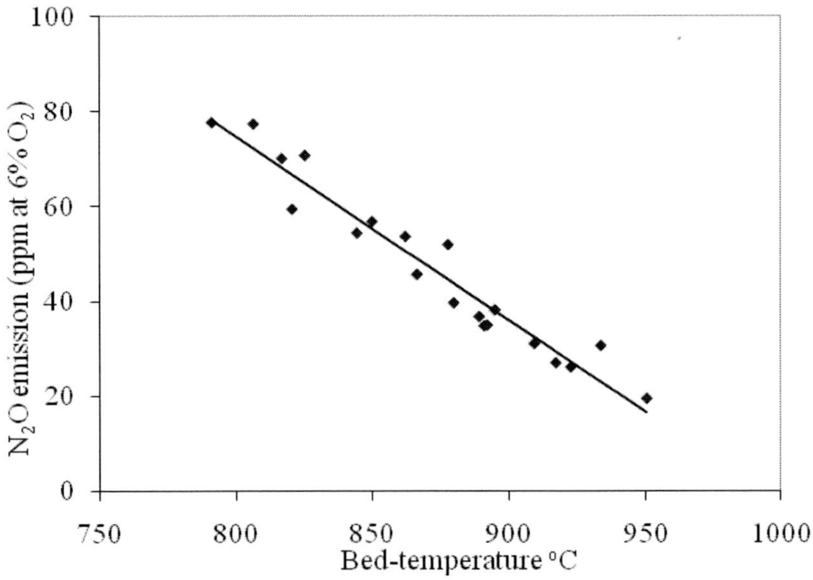

Figure 21. N_2O emission as a function of bed-temperature.

The pronounced increase of N_2O at high excess air can be explained as follow. As excess air increases, the convective cooling reduces the bed temperature, which in turn reduces H and OH radicals. When excess air increases, H radicals tend to be removed, leading to less N_2O destruction and hence higher N_2O emission as compared to the case of lower excess air. Figure 21 shows clearly the correlation of N_2O emission with the bed temperature.

CONCLUSIONS

Co-combustion tests of a high-moisture MSW with Thai lignite have been performed in a laboratory-scale fluidised bed to study the effects of MSW mass fraction on temperature and carbon combustion efficiency and on emissions of major nitrogenous species: NO, N_2O, HCN and NH_3. The results of the tests, with MSW fraction varied from 0% (or 100% lignite) to 40% at constant excess air of 40, 60, 80 and 100%, can be summarised as follows.

1. The addition of MSW to lignite causes the bed temperature to drop due to the much higher moisture and volatile contents of MSW in comparison to lignite.
2. Combustion in the freeboard is promoted by the higher volatile content of MSW.
3. Carbon combustion efficiency decreases slightly with the amount of MSW added, but becomes more significant with increasing emissions of CH_4 and CO as MSW increased beyond 30% because of moisture-inhibited hydrocarbon oxidation.
4. The reactivity of the MSW/lignite mixture appears to be enhanced at 20% MSW mass fraction, leading to rapid devolatilisation in the vicinity of the splash zone.
5. NO emission generally increases with increasing MSW, mainly due to increased formation of NH_3 and the less catalytic reduction of NO by char particles. However,

at 20% MSW, NO emission drops to a minimum for a given excess air because of less fuel particle available for the catalytic conversion of NH_3 to NO.

6. The emission of N_2O decreases slightly to a minimum at 20% MSW and then increases again as the fraction of MSW increases. The decrease is due mainly to the temperature drop while the increase may be attributed to the increase of CH_4 concentration and NO.
7. Unlike HCN, the emission of NH_3 can be fairly significant, particularly at low excess air conditions. Overall, adding MSW to lignite contributes to higher conversion of total fuel-N to NO, N_2O, HCN and NH_3, except at 20% MSW, because of higher volatiles in MSW.

In addition, the study of effect of MSW moisture content on the combustion and emission of NO and N_2O have been performed. Major findings from the test results can be summarised as follows.

a) The bed temperature in the case of 35%-moisture is higher than that for in case of 60% due to the difference of physical properties such as the percentage of fixed carbon and moisture content.
b) The combustion efficiency for the case of 35% waste moisture is higher than that for 60% moisture due to higher bed temperature at lower waste moisture. The synergistic effect of the co-combustion of lignite with the drier MSW reduces the emission of CO leading to increase in combustion efficiency.
c) The fuel mixture with low-moisture in waste gives higher fuel-N conversion to NO whereas the fuel-N conversion to N_2O is higher for higher moisture waste, particularly at high excess air.

REFERENCES

[1] Bowman, C.T., (1975) "Kinetics of pollutant formation and destruction in combustion" *Prog. Energy Combust. Sci.* 1, 33-45.

[2] Bromly, J. H., Branes, F.J., Little L.H., (1988) "The effect of low levels of CO, H2 and hydrocarbons on NO_2/NO ratios in heated gases" *J. Inst. Energy.* 89, 89-97.

[3] Chen, Z., Lin, M., Ignowski, J., Kelly, B., Linjewile, T. M. Agarwal, P. K., (2001) "Mathematical modeling of fluidized bed combustion: N_2O and NO_x emissions from the combustion of char" *Fuel.* 80, 1259-1272.

[4] Cliffe, K.R., Patumsawad, S., (2001) "Co-combustion of waste from olive oil production with coal in a fluidized bed" *Waste Manage. 21, 49–53.*

[5] Desroches-Ducarne, E., Christophe, J., Marty, E., Martin, G., Delfosse, L., (1998) "Modelling of gaseous pollutants emissions in circulating fluidized bed combustion of municipal refuse" *Fuel. 77,* 1399–410.

[6] Desroches-Ducarne, E., Marty, E., Martin, G., (1997) "The use of a paper/wood/plastics mixing as a model waste to study the incineration of municipal solid waste in fluidized bed" *Proc., the International Conference on Fluidized Bed Combustion. ASME.* pp. 769-777.

[7] Frankenhaeuser, M., Hitunen, M., Manninen, H., Palonen, J., Ruuskanen, J., Vartiainen, T., (1994) "Emissions from cocombustion of used packaging with peat and coal" *Chemosphere. 29,* 2057–66.

[8] Furusawa, T., Honda, T., Takano, J., Kunii, D., (1978) "Abatement of nitric oxide emission in fluidized bed combustion of coal" *J Chem Eng Jpn. 11,* 377-384.

[9] Hampartsoumian, E., Gibbs, B.M., (1982) "The influence of NH3 addition on emissions from a coal fired fluidized bed combustor" *Nineteenth Symposium on Combustion. pp.* 1253-1262.

[10] Jahkola, A., Lu, Y, Hippinen, I., (1991) "The emission and reduction of NOx and N2O in PFB-combustion of peat and coal" *Proc., the International Conference on Fluidized Bed Combustion. ASME. pp.* 725-730.

[11] Johnsson, J. E., (1994) "Formation and Reduction of Nitrogen Oxides in Fluidized Bed Combustion" *Fuel. 73,* 1398-1451.

[12] Leckner, B., Gustavsson, L., (1991) "Reduction of N_2O by gas injection in CFB boilers" *J. Inst. Energy. 64,* 176-182.

[13] Loffler, G., Andahazy, D., Wartha, C., Wunter, F., Hofbauer, H., (2001) "NO_x and N_2O formation mechanisms – a detailed chemical kinetic modeling study on a single fuel particle in a stationary fluidized bed" *Proc., the 16th International Conference on Fluidized Bed Combustion.*

[14] Manninen, H., Peltola, K., Ruuskanen, J., (1997) "Co-combustion of refuse-derived and packaging-derived fuels (RDF and PDF) with conventional fuels" *Waste Manage Res. 15,* 137-147.

[15] McGowin, C.R., Howe, W.C., (1994) "Alternate fuel co-firing in fluidized bed boilers" *Proceedings of The Industrial Power Conference. ASME. 24,* 155-70.

[16] McGowin, C.R., Huges, E.E., (1992) "Efficient and economical energy recovery from waste by co-firing with coal" In: Khan M.R. (Eds), *Clean energy from waste and coal,* ASME., Washington, DC., pp. 14–26.

[17] Philippek, C., Werther, J., (1997) "Co-combustion of wet sewage sludge in a coal-fired circulating fluidized bed combustor" *J. Inst. Energy. 70,* 141-150.

[18] Sorum, L., Gronli, M.G., Hustad, J.E., (2001) "Pyrolysis characteristics and kinetics of municipal solid wastes" *Fuel. 80,* 1217-1227.

[19] Suksankraisorn K., Patumsawad S., Vallikul P., Fungtammasan B. and Accary A., (2002) "Co-combustion of lignite and municipal solid waste in a fluidized bed: emissions of nitrogenous species" *27th International Technical Conference on Coal Utilization & Fuel System, Florida, USA, pp.* 313-324.

[20] Wargadalam, V.J., Loffler, G., Winter, F., Hofbauer, H., (2000) "Homogeneous formation of NO and N2O from the oxidation of HCN and NH3 at 600-1000°C" *Combust Flame. 120,* 465-478.

[21] Wheele,r P.A., Patel, M., Painter, A., (1995) "Fluidized bed combustion of municipal solid waste" *Proceedings of the International Conference on Fluidized Bed Combustion. ASME.* 597-607.

[22] Wartha, C., Winter, F., Hofbauer, H., (2000) "The trade-off between N_2, NO, and N_2O under fluidized bed combustor conditions" *Journal of Energy Resources Technology. ASME. 122,* 94-100.

In: Municipal Solid Waste
Editors: G. Konno and C. Machado

ISBN 978-1-62100-257-4

Chapter 5

AEROBIC SOLID-STATE PROCESSES TO MITIGATE GREENHOUSES GAS EMISSION IN MUNICIPAL SOLID WASTE MANAGEMENT

Barbara Scaglia**, *Valentina Orzi and Fabrizio Adani
Gruppo RICICLA- Dipartimento di Produzione Vegetale – DiProVe
Università degli Studi di Milano, Milan, Italy

ABSTRACT

With increasing the greenhouse gases (GHGs) gas emissions into the atmosphere and the possible consequence for the global climate change, there has been a growing interest in develop efficacy management strategies of reducing anthropogenic GHGs emissions.

Municipal solid waste (MSW) management plays an important role in the GHGs emission. The contribution of the landfilled MSW to GHGs it was reported to be of 5% of the total GHGs emitted and of the 13% of the methane from anthropogenic fonts (USEPA, 2003; IPCC, 2006).

The compute of the total GHG emissions is determined quantifying the GHG gases from all human activity as credits and debits. Debits are the GHG emitted in the atmosphere; on the other hand, the credits are the storage on C in stable forms (C-based systems) (IPCC, 2006; Brown et al., 2008).

Because the production of CH_4 *vs* CO_2 during MSW degradation depends by aerobic *vs* anaerobic conditions, the possibility to manage MSW organic fraction by aerobic biological process, i.e. composting and mechanical biological treatment, avoiding methane production in landfill, reduce GHGs emission due to MSW management.

The potential reduction of the methane emission and the possibility to store C in landfill and soil under recalcitrant forms, is extensively discussed in this chapter, reporting both literature and new data.

* Tel: +39-02- 50316546, Fax: +39-02-50316521, e-mail: barbara.scaglia@unimi.it.

1. Introduction

1. Greenhouse Gases Emission and Climate Change

Anthropogenic emission of greenhouse gases was rapidly increased with the Industrial Revolution and this event has caused a world interest in identifying strategies of reducing the rate of gaseous emission.

The intergovernmental Panel on Climate Change shows that from 1850 and 1998 the emission from terrestrial ecosystem was about half of fossil fuel combustion.

Municipal solid waste management plays an important role in the greenhouse gases (GHGs) emission. The contribution of the landfilled MSW to GHGs, it was reported to be of 5% of the total GHGs emitted and the 13% of methane from anthropogenic fonts (USEPA, 2003; IPCC, 2006). As consequence of that, MSW management needs to be revisited in order to get sustainable approach able to reduce this contribution.

2. Municipal Solid Wastes Generation

Municipal Solid Wastes are defined as the waste fraction collected by a municipality. It concerns waste from households, small business, office buildings and institutions such as schools, hospitals, government buildings, waste from parks and street cleaning (EUROSTAT, 2006).

Table 1. Composition of MSW in the world (Scaglia and Adani, 2008a)

Fraction	UE[†]	African cities[‡]	USA[§]	Asia (urban zone)[¶]
	Content (% by weight)			
Paper	25-35	0-12.9	35.2	4.2-19
Plastic	7-10	0-6.3	11.3	4-19
Metals	3.5-7	0-2.6	8	0.1-6
Glass	5-10	0-1.9	5.3	0-3
Ceramic	1-2	-	-	-
Yard Waste	10-15	-	12.1	-
Hazardous Waste	10-15	-	-	-
Rubber, Leather And Textiles	-	0-3	7.4	0-9
Wood	-	-	5.8	-
Food, Vegetable And Fruits	25-35	-	11.7	-
Other	-	-	3.4	2-37.7
Biodegradable Fraction	-	55.8-94	-	39-74
Organic fraction[#]	60-90	55.8-100	72.1	43.2-100

[†] Bidlingmaier et al., 2004.

[‡] Asomani-Boateng and Haight, 1999.

[§] USEPA, 2003.

[¶] Zurbrügg, 2002.

[#] Organic fraction=paper[+]rubber[+]leather[+]textiles[+]wood[+]food[+]vegetable[+]fruits[+]biodegradable fraction.

[+] yard waste.

Generally the MSW composition takes in: (1) food waste, (2) garden (yard) and park wastes, (3) paper and cardboard, (4) wood, (5) textiles, (6) nappies (disposable diapers), (7) rubber and leather, (8) plastics, (9) metal, (10) glass (and pottery and china) and (11) other (e.g. ash, dirt, dust, soil, electronic waste).

The composition and the volume of disposed wastes vary nationally and regionally in relation to human activity and to the quantity and type of products that communities consume (Taylor and Allen, 2001) (Table 1).

2.1. MSW Landfill Disposal

The landfill represents the most common way of municipal solid waste (MSW) disposal. Data of 2006 (IPCC, 2006), reported that about 67 %, 69% 61% 63% and 75 % of the total MSW generated are landfilled in Europe, Africa, America, Asia and Oceania, respectively. In the landfill, the biodegradation of the organic fractions is the major mechanism that governs biological pollution (odors, biogas and leachate) (Derham, 1995; Driessen et al., 1995; Westlake, 1997; Read et al., 2001; Visvanathan, 2005). The uncontrolled productions of landfill gas (CH_4, CO_2 and trace of non-methane volatile organic carbons) lead to ozone depletion and contribute to global warming effect (Bogner and Metthews, 2003). The composition of biogas depends on the landfill phases (Bockers and Steinberg, 2005). In the Phase I (aerobic phase), the gases are mainly nitrogen and oxygen. The microorganism metabolism takes up the amount of CO_2 concentration. During the Phase II (acidogenic phase) the oxygen is consumed completely; the emitted gases are hydrogen and CO_2 while nitrogen remains present. Phase II is instable and the process can reverse back if the oxygen contacts the waste.

During Phase III, the chemical condition for the growth of methanogenic community is created. Therefore, the CH_4 production increases to reach a level of 50-60% v/v (Bockers and Steinberg, 2005). At the same time, the concentration of CO_2 decreases from 70% v/v to 40% v/v. In this phase the production of biogas is large but lower than Phase III.

The production during this phase remains constant as both amount and composition (CH_4:CO_2 ratio is 1.25) (Bockers and Steinberg, 2005).

Phase V is the long-term phase in which most of the gas emission occurs with pressure inside the landfill higher than atmospheric pressure. The air infiltration in the landfill determines the re-start of aerobic microorganism metabolism and the oxygen oxides the methane inside the landfill body with a progressiving reduction to 0% v/v (Bozkurt et al. 1999; Revans et al. 1999; Kjeldsen et al. 2002; Scaglia and Adani, 2008a).

2.2. Ghgs Emission from Landfill

Landfills are the first cause of emission of GHGs (IPCC, 2006) in the MSW management system. GHGs emitted by landfill consisted of CH_4, CO_2, N_2O. The landfill gas is composed of methane (50%), carbon dioxide (45%), nitrogen (5%) and hydrogen sulphide (<1%) (Scaglia and Adani, 2008a). Methane has an infrared activity 21 times that of CO_2 (Lelieved et al., 1998). Because the production of CH_4 *vs* CO_2 depends by aerobic *vs* anaerobic conditions, the possibility to manage MSW organic fraction degradation under aerobic condition instead of anaerobic ones such as occurs in landfill, allows avoiding methane production, reducing greenhouses impact of MSW management.

Considering the total anthropogenic production of GHGs, the MSW contribution is estimated in different values by different source (IPCC, 2006) (Table 2) depending by the difficult to make the inventory of GHGs state emissions for this sector. Some Authors reported that of the total anthropogenic methane emissions on Earth, calculated for the year 2000 to be of 282.6 x 10^6 Mg (USEPA, 2003) (Table 2), the 13% (36.2 x 10^6 Mg) was due to landfill emissions (Table 2). The GHGs estimation considered only the CH_4 production because the CO_2 from biogenic fractions is not computed in the greenhouse effect inventory (IPCC, 2006) and the other compounds are in less quantity.

The emission of methane from landfill has been determined considering both theoretical and experimental approaches. The theoretical approach supposes that the organic fraction of MSW decomposed under optimized condition so that maximum biodegradation rate is expected (Bogner and Matthews, 2003). When the whole MSW is considered, the literature reports theoretical CH_4 productions for different countries of 0.18 kg CH_4 kg^{-1} MSW on a wet weight basis for U.S. and 0.25-0.47 kg CH_4 kg^{-1} MSW on a wet weight basis for Germany and Italy) (Halvadakis et al., 1983; Bogner and Spokas, 1993; Scaglia and Adani, 2008a).

However, the real landfill condition does not allow the highest methane production to be produced, giving low CH_4 emission, i.e. 0.036 kg CH_4 kg^{-1} dry solid waste (Bogner and Matthews, 2003).

Data referred to landfills indicated different methane production ranging from 0.1 to 1.0 Mg CH_4 ha^{-1} d^{-1} (Nozhevnikova et al., 1993; Hovde et al., 1995; Borjesson, 1996; Czepiel et al., 1996; Mosher et al., 1999; Galle et al., 2001; Morris et al., 2001). For a landfill in South Italy, it was reported a biogas production ranged from 7,520 to 10,154 m^3 h^{-1} (for an area of 47 Ha and 41.5 Mg of MSW treated per day), the 50-60% of which were emitted directly into the atmosphere (Aronica et al., 2009).

The type of site in which waste is disposed also significantly influences CH_4 generation and the different time-scale of release (Figure 1). There are three types of disposal sites: open dump, dry tomb and containment landfill. Obviously dry tomb but above all containment landfill reduces greenhouse gases emission (Scaglia and Adani, 2008).

Table 2. Global anthropogenic methane emissions (Scaglia and Adani, 2008a)

Fraction	Anthropogenic methane emission (% of the total amount)
	USEPA, 2003
Enteric fermentation	28
Natural gas	15
Solid waste	13
Rice	11
Waste water	10
Coal	8
Biomass burning	5
Bio- fuel combustion	4
Manure	4
Oil	1
Fuel stat and mobile	1

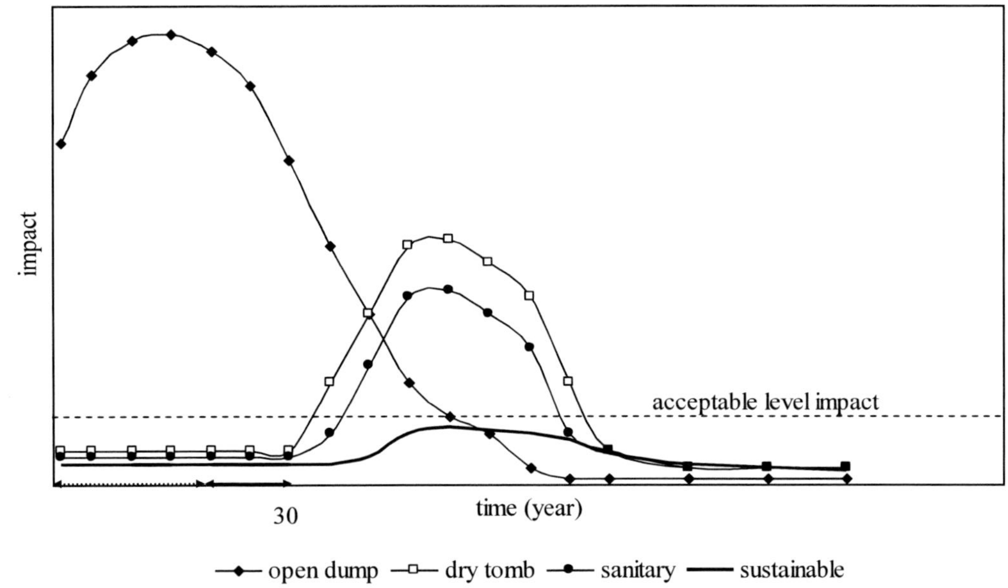

Figure 1. Impact level for different landfill type (Visvanathan, 2005, modified).

2.3. The Reduction of Biogas Emission from Landfilled MSW

The change of climate as consequence of the increase in the atmosphere of the greenhouse gases, suggests making any efforts to reduce the world GHGs emissions.

Kyoto protocol requires the reduction of GHGs production of all human activities sectors, including the MSW management (IPCC, 2006).

The compute of the total GHGs emissions is determined quantifying the GHG gases from all human activity as credits and debits. Debits are the GHG emitted in the atmosphere; on the other hand the credits are the storage on C in stable forms (C-based systems) (IPCC, 2006; Brown et al., 2008).

Different waste-management systems contribute to GHGs reduction. The benefits achieved can be direct: when they determined a direct reduction of GHG emission, or indirect: when their applications avoid significant GHGs generation (IPCC, 2006).

For example, the recovery of methane from existing landfill to produce energy is an efficient direct GHGs mitigation-way that can be applied to existing landfills.

On the other hand, since methane emission from landfills takes long time before to be completed (20-30 years), the biological pre-treatments of MSW are considered complementary GHGs mitigation procedures.

The reduction of the content of the organic fraction of MSW (OFMSW) to be landfilled can be obtained by three different approaches: i) MSW incineration to produce energy; ii) source separated collection of OFMSW to produce compost; iii) mechanical-biological treatment (MBT) of MSW to produce a compost-like material to be landfilled (Table 3):

1. the incineration provides significant renewable energy benefits avoiding the use of fossil fuel. On the other hand the incineration determined the release of GHGs (40 x 10^3 Mg CO_2-eq y^{-1}) (IPCC, 2006) because of the burning of the non biogenic

fractions (e.g. plastics, synthetic textile materials). Moreover, no significant GHG credits are get because of the loss of the whole organic matter content that cannot be employed to form C sink.

2. The source-separated collection of OFMSW determined an indirect effect on GHG emissions, avoiding the methane production in landfill, when it is transformed in compost. Moreover, the application of compost in agriculture allows soil to act as C-sink for a medium-long time period, restoring at the same time soil carbon content. At the same time the nutrients contained in the compost allows to save the energy needs to produce chemical fertilizers (IPCC, 2006; Brown et al., 2008)
3. The debits of composting about GHGs, depends by both the energy consumption (electrical energy and fossil fuel) and the composting process (pre- and post-treatments, aerated composting, static piles, etc.) and biological gas emissions (CH_4 and N_2O). Nevertheless the debits due to the both energy utilization and the gas emissions, are retained to be minimal (IPCC, 2006; Amlinger and Peyr, 2008; Brown et al., 2008).
4. The MBT is a composting-like process that use as starting material unsorted or residual MSW. Similarly to the composting process, the MBT process reduces the methane emission coming from the degradation of easily degradable organic fraction as this fraction, under aerobic condition, is transformed in CO_2, avoiding the CH_4 production. The no degraded OM represents the more recalcitrant fraction that could be preserved for long time in the landfill, acting, this latter, as carbon sink (Barlaz, 2006; Kelly et al., 2006; IPCC, 2006; Ximenes et al., 2008).

Table 3. Qualitative effects on GHGs credit and debit for different MSW managements (IPCC, 2006; Brown et al., 2008)

MSW managements	GHGs credits			GHGs debits
	Indirect	Direct	C based system (C sink)†	Debits
Incineration	Fossil fuel offsets	Renewable energy use	No (<1%)†	Minimal: Incineration of non biogenic fractions
Composting	Avoid GHG production in landfill	Compost production (fossil fuel offsets to produce amendments)	Yes (soil C sink) (15-50%)†	Minimal: 1.Energy consumption 2.CH_4 and NO_2 emissions
MBT	Avoid GHG production in landfill	MSW compost-like production	Yes (landfill C sink) (>50%)†	Minimal: 1. Energy consumption 2. CH_4 and NO_2 emissions

† IPCC, 2006.

3. Discussion

3.1. The Chemical Composition of the Organic Matter Contained in the MSW

The knowledge of the composition of the MSW is important as MSW contains organic fraction that, depending by its composition and environmental conditions can produce GHGs. The OM of MSW can be divided in biogenic and no-biogenic fraction. The biogenic fraction, can be defined as the organic matter produced in natural processes by living organisms but not fossilised or derived from fossil resources (CEN/TC 343). This fraction is represented by food waste, garden waste, paper, wood, textile etc. and from a chemical point of view it consist of lipids, carbohydrates, proteins and lignin (Picheler and Kögel-Knaber, 2000).

The composition of MSW varies considering different origins (Table 4), although carbohydrates represents the most abundant fraction (Table 4) followed by the lipid fractions. The ligno-humic fraction is the less abundant fractions characterizing MSW.

The no-biogenic fraction, consist in fossil-derived organic matter: plastics, rubber and petroleum-derived chemicals.

The biogenic fraction is the responsible for the production of GHGs when MSW is landfilled. Therefore, the quantity and quality of this fraction in MSW determine the extent to which MSW produce GHGs.

Composting and MBT are MSW treatments based on a solid-state aerobic biological process able to modify both quantitatively and qualitatively the OM of MSW and as consequence the potentiality to produce GHGs.

In this way it is important to understand how these processes, from a biochemical point of view, affect the chemistry of the OM of MSW.

Table 4. Biogenic organic fraction composition of MSW

OM fraction	Italian OM from separated collection†	German-MSW‡	Italian-MSW†	Residential food from USA§
	$g\ kg^{-1}$ VS			
Lipid	209.3	138	141.0	-
Non –cellulosic carbohydrates	-	190	-	-
Hemicelluloses	300.2	-	436.7	125
Proteins	-	71	-	-
Cellulose	204.5	257	242.8	631
Lignin	285.9	35	285.9	140
not indefinable organic matter	-	308	-	-

† DiProVe data.

‡ Picheler and Kögel-Knaber, 2000.

§ Barlaz, 2006.

3.1.1. Biodegradability and Recalcitrance of OM: Definitions

The solid-state aerobic process is a bio-oxidative process that determines the biodegradation of the labile organic fractions and the preservation/modification of the more recalcitrance molecules, giving a biologically stable product.

The definition of biodegradability and of recalcitrance is fundamentals to evaluate the efficacy of solid-state biological process to reduce GHGs emission and/or to act as carbon sink.

A recent definition of biodegradability can be given as in the following.

Biodegradable material: capable of undergoing biological anaerobic or aerobic decomposition under conditions naturally occurring in the biosphere (Landfill Directive 1999/31/EC). The aerobic biodegradation of organic matter determine the release of CO_2 while the anaerobic condition produce CH_4 and CO_2.

Theoretically, all biomasses are biodegradable (European Committee for Standardization, 2006). The time-scale considered and the environmental conditions occurred during biomass degradation, influence the rate of OM degradation (Bogner and Metthews, 2003). As consequence of that, the biodegradability of a biomass is of difficult definition. One approach is to define both time-scale and conditions during which degradation occurs.

Recalcitrance can be defined as the natural resistance of biomolecules to microbial and enzymatic deconstruction (Himmel et al., 2007). The biochemical recalcitrance is linked to the both chemical and molecular conformation of biomolecules: aromatic and alkyl compounds such as lignin, cutin, suberin and tannins, are resistant to biodegradation. Generally to the aromatic compound (lignin) is attributed the major chemical recalcitrance, but recent studies have suggested that, also, condensed tannins and polyphenols, cutin and suberin are recalcitrant (Rasse et al., 2005). Such as for the biodegradability definition, also the recalcitrance is affected by both the time-scale considered and the environmental condition occurred during biological process.

Biological, chemical and physical parameters, characterizing organic compounds, and the environment conditions occurred during degradation, determined, also, the recalcitrance *vs* biodegradability balance of the organic matter (OM). For example, low nutrient contents, an uncorrected nutrient ratio, unsuitable environmental conditions, the presence of toxic substances, and the no availability of the organic compounds, could slow down degradation processes.

Considering the solid-state biological process, the recalcitrance-degradability balance depends by the follows conditions:

1. *Environmental conditions.* The process conditions are not favourable for microbial metabolism. The more frequently adverse conditions are the pH (low values), the moisture content (lowest/highest contents), the too high concentration of toxic microbial metabolism products (i.e. volatile fatty acids). The recalcitrance character of the OM in these cases is transitorily, i.e. the disappearance of these adverse conditions allows the biodegradation of OM.
2. *Chemical recalcitrance.* The chemical recalcitrance of OM referred to physic-chemical characteristics of OM (kinetic degradations). Otherwise the decrease of the biodegradability of the OM fraction depends by the presence of another component. For example the biodegradation of cellulose is reduced by the lignin presence

because of physical and chemical inhibitions (Barlaz, 2006; Komilis et al., 2006; Ximenes et al., 2008). In the vegetal materials, the lignin present among the cellulose fibrils caused a decreased of the available surface for the microbial attach to cellulose. Moreover, the adsorption of extracellular enzymes (cellulosic enzymes) on lignin, increase the cellulose recalcitrance.

3. *Physical recalcitrance.* The physical properties of the substrate (degree of polymerization and cross-link of the organic molecules) affect the specific surface available for biological reaction and, the quantity and quality of the porosity (micro- *vs* macro porosity) influence the accessibility of the OM structures to enzymes (El Fadel et al., 1997).

Studies conducted in our laboratories have demonstrated that not only the chemical composition but, also, the physical structure affects the OM degradation. Most of the hydrolytic enzymes have a molecular weight of 20-25 kD, corresponding to a size of about 4 nm, that hardly penetrate into micropores and small mesopores of the OM structures, so that their activities are confined only to a portion of the total surface (Zimmerman et al., 2004; Chesson, 1997).

3.2. Solid State Aerobic Process: Biochemistry Basis and Degradation Processes

3.2.1. Biochemical Aspects of Aerobic Process

The solid-state biological process consists in a series of exothermic oxidative biological reactions that determine the oxidation of labile organic fractions and the concentration of the more recalcitrant fractions.

The biochemical reactions can be synthesized as:

$$[CH_2O]^{+} O_2 \rightarrow CO_2^{+} H_2O^{+} \text{energy}^{+} \text{recalcitrant-C} \quad (1)$$

The equation 1 resumes a series of more complexes reactions with the production of intermediate metabolic products that are the substrate for the following anabolic and catabolic reactions.

The aerobic metabolic reactions are regulated by variables that influence the substrate availability and concentration (i.e. microbiological activity) and by variables that influence directly kinetics reaction (i.e. temperature, moisture, organic fraction characteristics) (Hamoda et al., 1998). The control of these variables is fundamental to perform the composting and MBT processes efficiently.

The aerobic biological process is constituted by a system of three phases: solid, liquid and gaseous.

The solid phase is the particle that is constituted of organic and inorganic fractions. The particle interacts with liquid phase that is adhering to its surface. The liquid phase constitutes the medium where microorganisms live and where the biological chemical reactions occurred. The organic fractions that are into the biofilm, are oxidize to build new microbial biomass, CO_2, H_2O and heat. The biofilm of the liquid phase, exchanges directly the gas (overall O_2 and CO_2) with the close gaseous phase.

The CO_2 and the O_2 fluxes are regulated by the concentration gradient. Moreover, the capability of O_2 flux penetrating into the particle depends by the internal resistance forces of the particle. Considering this, the biofilm's layer is subdivided in two strata: a thin outside aerobic strata and a thick inside anaerobic strata. In the anaerobic strata the insoluble organic fractions after enzymatic reaction is hydrolyzed in soluble organic fractions that are transformed in products by anaerobic metabolism (e.g fatty acids, etc..). Both the hydrolyzed organic fraction and the products of the anaerobic metabolism flux to the biofilm aerobic layer where are oxidize to CO_2 and H_2O.

Fractions that do not underwent to biological reaction, i.e. recalcitrant molecules, are preserved.

3.2.2. The Conservation/Modification of the Organic Matter During Solid-State Biological Process

The evolution of organic substances during solid-state aerobic process implies both quantitative and qualitative changes. During the process, the organic substances content are partially degraded by microorganisms activity. The labile organic fractions (i.e. simple lipids, proteins, cellulose and hemicelluloses) are degraded (Adani et al., 1995; Adani et al., 1997). Otherwise, more recalcitrant molecules are preserved and concentrated (Kögel-Knaber et al., 1992). Modification of re-calcitrance molecules, also, occurs during biological process. The process of concentration and preservation of recalcitrant molecules is known as the humification process (Adani et al., 1998). Humus represents a "dark brown or black amorphous material characterized by a large surface area, high charge density, high affinity for water. The main routes for humic substances formation include different theories (Stevenson, 1994). These theories hypothesize that biomacromolecules first break up into small constituents, and then subsequently recombine to give more complex organic matter (Hedges and Oades 1997). On the other hand many studies conducted more recently (e.g. Kögel-Knabner et al. 1992; Hedges and Keil 1995; Adani et al., 1995, 1997, 2006 and 2007) indicated that it is the preservation and modification of recalcitrant molecules that provides the humification pathway. On reviewing the origin of soil organic matter, Kögel-Knabner (2002) indicated humification to be the prolonged stabilization of organic substances, countering biodegradation, and to involve plant biopolymers such as lignin, cutin and suberin. Preservation of biomacromolecules is due to biochemical recalcitrance caused by the inherent property of the molecular structure (Kögel-Knabner, 2002). Therefore it can be concluded that humification consists in the preservation of recalcitrant bio-macromolecules (Kelleher, 2006).

According to this concept the HS are a complex of distinct of faunal, microbial and plant biopolymers and proteins, lignin, polysaccharides and aliphatic polymers (Kelleher, 2006; Lorenz et al., 2007). Therefore, humification represents more correctly, the preservation of molecules because of their biological recalcitrance.

3.3. The Characterization of the Final Products of the Biological Process: Biodegradability vs Recalcitrance of MSW Organic Matter

The measure of the efficacy of the aerobic process consisted in a direct or indirect measurement of the degree of decomposition of the easily degradable OM *vs* the presence of recalcitrant OM. As the success of aerobic biological processes *vs* GHGs emission is the avoiding of CH_4 emission in landfill and the possibility to use compost and biostabilized material as C-sink, this evaluation became fundamentals for the success of these strategies in the reducing GHGs emissions.

In the following sub-chapters the most used and validated method used to measure the degree of decomposition *vs* recalcitrance of the treated MSW, are discussed and used in the successive chapters to measure the effect of aerobic biological processes on MSW GHGs reduction.

3.3.1. Biological Approach: Respirometric Activity Measurement

The biological stability determines the extent to which readily biodegradable organic matter has decomposed (Lasaridi and Stentiford, 1998). It identifies the actual point reached in the decomposition process and represents a gradation on a recognized scale of values, which thus enable comparison of the process of decomposition (Lasaridi and Stentiford, 1998). Therefore knowing the degree of biological stability possessed by the organic matter contained in wastes is important (Müller et al., 1998; Science and engineering of composting, 1993; Adani et al., 2004a).

Many indicators and indexes have been proposed in the past to measure the biological stability of wastes (Komilis et al., 1999; El Fadel et al., 2002; Fricke et al., 2005; Robinson et al., 2005; Kelly et al., 2006). Nevertheless, the most used and accepted methods to measure the biological stability are the respirometric methods (Iannotti et al., 1992; U.S. Composting Council, 1997). A detailed review of the different respirometric methods can be found in Barrena Gómez et al. (2006). DiProVe approach consisted in the use of a Dynamic Respiration Index to measure the biological stability degree of wastes (Scaglia and Adani, 2008a). Method set up is nowadays both a national (UNI, 2006) and a European Standard method (CEN, 2006).

The oxygen uptake rate (OUR) is a direct measure of the microbial activity and it is highest at the start of the biological process when large amount of easily biodegradable fractions are present (Table 5), becoming low at the end of the process when recalcitrant molecules became more abundant than degradable ones (Table 5).

3.3.2. Chemical Approach: Total Organic Matter Contents and Macromolecular Composition

The solid-state aerobic process causes a qualitative/quantitative modification of the OM of MSW. Therefore the study of the evolution of the OM of MSW allows evaluating the efficacy of process and the characteristics of final products. This can be done by different approaches.

Table 5. Biological stability evolution during composting and MBT processes

Process	Treatment period[†]	Reference		Respiration activity
composting				
	0	Said Pullicino et al., 2007	SOUR method (mg O_2 gVS^{-1} h^{-1})	13.1±0.6
	13 d			27.5±2.5
	25 d			25.6±1.4
	90 d			6.3±0.5
	250 d			1.9±0.3
	0	Scaglia et al., 2007		9
	21 d			4.94
	90 d			2.7
	0	DiProVe data	DRI (mgO_2 kgVS^{-1}h^{-1})	1815
	17 d			830
	42 d			389
	90 d			207
MBT				
	Initial	Scaglia and Adani, 2008b	DRI (mgO_2 kg VS^{-1}h^{-1})	2817±1184 n=8 processes
	intermediate			1532±790 n=3 processes
	Final			827±182 n=8 processes
	0	Scaglia et al., 2007	SOUR method (mg O_2gVS^{-1})	8.58
	45 d			5.72
	75 d			2.58
	0	Zach et al., 2000	AT_4 (mgO_2 gdm^{-1} 96h^{-1})	32
	5 w			28
	9 w			24
	20 w			2.5
	0	Lornage et al., 2007		82.9
	4 w			58.7/57.6
	8 w			35.3
	16 w			20.5
	25 w			16
	3	Münnich et al., 2006		13.1
	6 m			6.9
	12 m			1.7

[†] d= days, w= weeks, m= months.

THE VOLATILE SOLID CONTENT

The volatile solid content is a simple parameter used to evaluate the performance of biological processes. The VS content can be assumed for the MSW to be equal to the OM. Therefore, in the MSW science it can be used indifferently both the term OM and VS.

The VS content is a low cost and quick analysis. During biological process, the relative (g kg^{-1}) and absolute (weight units) VS contents decrease because of degradation processes (Table 6). Although this method is simple, it does not give any information on the quality of the residual organic matter. Moreover it is not able to discriminate between biogenic and no-biogenic fraction (Table 6).

Table 6. Relative and absolute volatile solid trend during solid state aerobic processes

Process†	Treatment period‡	VS	
		Relative value (g kg^{-1} dm)	Absolute value (g)
C	0§	540	540
	16 w§	310	207
	55 w§	260	162
	Reduction§ (g kg^{-1})	*518.5*	*700*
MBT	0#	707±105	707
	Intermediate#	599±107	437
	End (3-4 w)#	579 ± 88	403
	Reduction# (g kg^{-1})	*181*	*430*

† C: composting process; MBT: mechanical biological process.

‡ w=week.

§ Pichler and Kögel-Knaber, 2000.

Scaglia and Adani, 2008b. The data are the mean of 5 different MBT processes.

MACROMOLECULAR COMPOSITION OF OM: WET ANALYSES

The fractionation of the organic matter in macromolecular classes can be useful to infer about the quality, i.e. degradability *vs* recalcitrance, of the OM of MSW. Literature reported different procedure to fractionate OM, all grouped under the denomination of wet chemical fractionation (Inbar et al., 1990; Van Soest, 1991; Pichler and Kögel-Knaber, 2000). Pichler and Kögel-Knaber (2000) divided the OM of MSW in five different fractions: no- cellulosic carbohydrates, cellulose, proteins, lipids, lignin and unidentified OM (i.e. plastics, leather, humic substances) (Table 7). The DiProVe (Adani et al., 1995) divided MSW into five fractions: I: lipids, wax, resins, proteins and fulvic acids; II: hydro soluble hemicelluloses and hydro soluble tannins, fulvic acids and soluble ash; III: cellulose; VI: lignin and humic acid (Table 7).

Table 7. Relative and absolute (bold letters) trend of macromolecular OM-fractions of MSW during solid state aerobic processes

Process†	Time	Refe-rences	Lipid	Non cellulose carbohy-drates	Cellu-lose	Pro-teins	Lignin	Unde-fined organic matter‡
					g kg^{-1} OM g			
MBT	0	Pichler	141	190	260	69	37	303
		et al.	141	190	260	69	37	303
	16 w	2000	127	100	150	81	79	463
			26	21	31	17	16	96
	55 w		120	80	100	95	84	521
			19	13	16	15	13	84
Reduction			*149*	*579*	*615*	*-377*	*-1270*	*-729*
(g kg^{-1})			*865.25*	*931.58*	*938.46*	*782.61*	*648.65*	*722.77*
			I§	II¶	III#	IV+		
C	0	Lhadi et	98.59	578.51	198.75	123.89		
		al. 2006	98.59	578.51	198.75	123.89		
	45d		102.24	503.03	169.10	225.46		
			46.92	230.82	77.59	103.46		
	165 d		102.16	428.77	120.90	348.17		
			24.29	101.95	28.75	82.78		
Reduction			*-3.62*	*258.80*	*391.70*	*-181*		
(g kg^{-1})			*753.63*	*823.77*	*855.36*	*331.80*		
MBT	0	DiProVe	141.0	436.7	242.8	179.3		
		data	141.0	436.7	242.8	179.3		
	Inter-		100	264.3	337.6	316.8		
	mediate		43.70	115.50	147.26	138.44		
	End (3-4		85.9	251.0	332.9	330.2		
	weeks)		34.61	101.5	134.16	133.00		
Reduction			*390*	*425*	*-370*	*-842*		
(g kg^{-1})			*754.54*	*768.38*	*447.40*	*258.23*		

† MBT: mechanical biological process; C: composting process.

‡ i.e. plastics, leather, humic substances partially.

§ I fraction: lipids, cere, resine, proteins and fulvic acids partially.

¶ II fraction: hydrosoluble hemicellulose and tannins, fulvic acid and soluble ashes partially.

III fraction: cellulose.

+ VI fraction: lignin and humic acid.

Table 7 reported the evolution of different macromolecular OM fractions during the aerobic solid state processes. Considering the relative content of different fractions, results shows the degradation of the more labile fractions (carbohydrates, lipids, proteins) and the concentrations of more recalcitrant ones.

The absolute OM fraction contents, resulted from the application of OM fractionation and the complete mass balance during aerobic biological process, showed that all organic fractions degraded during the solid state aerobic processes but at different rate. The carbohydrates showed the highest degradations rate followed by the lipid and proteins and later by lignin/ ligno-humic fractions. Aerobic process lead to the conservation of the recalcitrant molecules and degradation of the labile ones so that MSW OM resembled soil OM (Pichler and Kögel-Knaber, 2000).

The results of the macromolecular fractionation can be used to calculate index able to describe the degradation *vs* recalcitrance of OM (Table 8). Several Authors (Suflita et al., 1992; Revans et al., 1995; Pichelr and Kögel-Knaber, 2000) used the cellulose to lignin ratio (C/L) to describe the recalcitrance degree of OM. On the other hand, the DiProVe introduces the use of recalcitrant index (RI), proposed by Rovira e Vallejo (2002), to study plant residue degradation in soils, to evaluate the recalcitrance degree of MSW.

Table 8. Recalcitrance Index (RI) and Cellulose/Lignin ratio (C/L) measured for MSW during aerobic biological processes

Process typology[†]	Literature reference	Time	C/L (w/w)	RI[‡]
C	Pichler and Kögel-Knaber, 2000	0	7.05	-
		16 w	1.89	-
		55 w	1.19	-
C		0	5.42	-
		61 w	1.0	-
C	DiProVe Data	0	-	18.72
		11 d	-	26.01
		150 d	-	37.17
C		0 d	-	28.59
		90 d	-	56.06
MBT		0	2.21	17.93
		Intermediate	0.84	31.68
		End (3-4 weeks)	1.34	33.02

[†] C: composting process; MBT: mechanical biological process.
[‡] RI: recalcitrant index = (Recalcitrant OM/ Total OM) * 100.

Macromolecular Composition of OM: Spectroscopic Methods

The spectroscopic techniques give information about the evolution of OM during solid state aerobic processes. The spectroscopic analysis applied tipically used are the nuclear magnetic resonance (NMR) and the infrared spectroscopic (IR), which allow to obtain qualitative and semi-quantitative informations of organic substances (Sparks, 1998; Pichler and Kögel-Knaber, 2000).

3.4. Mechanical Biological Treatment and Composting *Vs* Ghgs Emission Reduction

3.4.1. Mechanical Biological Treatment and Composting Processes

Mechanical Biological Treatment (MBT)

The Mechanical Biological Treatment (MBT) refers to a process consisting in the mechanical fractionation of MSW into a wet (mostly organic) and a dry fraction (mostly plastic and paper) and in the successive biological treatment of the wet fraction (Wiemer and Kern, 1995; Adani et al., 2000; Calcaterra et al., 2000). In the last 15 years these kinds of processes found a wide diffusion in Europe (e.g. Austria, Germany and Italy).

The biologically treated wet fraction is known as biostabilized organic or compost-like fraction (Wiemer and Kern, 1995) and its fate is the landfill or its usage as organic amendment for particular uses (e.g. landfill and land restoration). On the other hand the dry fraction is landfilled or used to produce solid refuse fuel (SRF) (Leikam and Stegman, 2004). More recently mechanical biological processes assumed different forms and have been classified on the basis of the presence or absence of physical separation of waste into different streams before the biological treatment. Therefore it was defined the mechanical-biological treatment with separate fluxes (MBTSF) (Adani et al., 2000) and the mechanical-biological treatment with unique flux (MBTUF) (Calcaterra et al., 2000).

The MBTSF (Figure 2) consists in a mechanical pre-treatment of the waste (sieving) giving a wet (Ø $<$ 50-80 mm; OM $>$ 80 % w/w; moisture $>$ 50 % v/v) and a dry (Ø $>$ 50-80 mm; OM $<$ 25 %; moisture $<$ 25-28 %) fraction, and the successive biological treatment of the wet fraction. This treatment went on for several weeks using an aerobic biological treatment in order to reduce the content of putrescible organic fraction leading to a high degree of biological stability (biostabilized product). The length of the biological treatment can vary, so that a biostabilized product shows different degree of biological stability.

In the MBTUF process (Figure 3) the entire waste after a bland mechanical treatment used to reduce waste size, is processed by a biological process. Depending on the process management (air, temperature and moisture management) fast and long biological treatments can be indicated (Adani et al., 2002). A fast biological process gives a well-dried material with a high calorific value and a low degree of biological stability (Adani et al., 2002). In this case, the fate of the biodried material is the production, after refining, of SRF. Long biological process gives a product characterized by low calorific value (depending on the process) and high biological stability (Adani et al., 2002; Forte et al., 2003) (Figure 3).

A particular application of the MBTU is to perform the process directly in landfill sites allowing a low-cost biostabilization technology. This option of pre-treatment leads to reducing the plant cost and preserving high process performance (Adani et al., 2000; Forte et al., 2003). Moreover, the modularity of system and the low cost favoured their application in developing and emerging countries. More details can be found in a recent publication (Scaglia and Adani, 2008a).

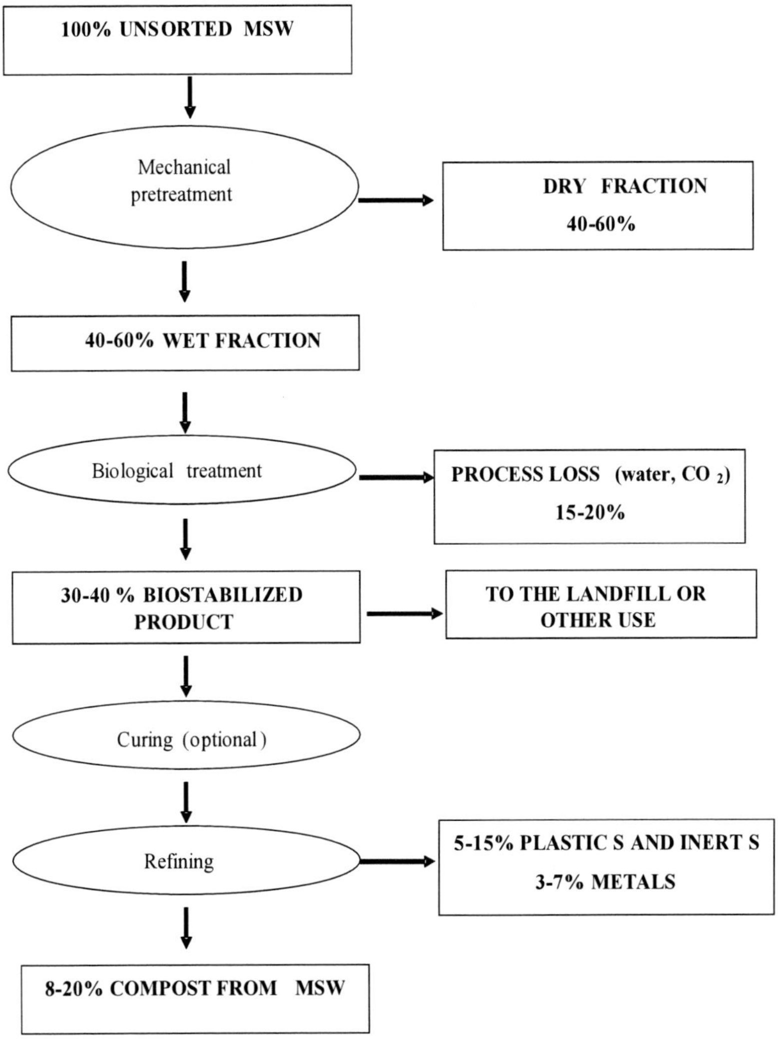

Figure 2. MBTSF mass balance scheme.

Composting Process

The term composting defines a solid-state aerobic treatment that produces compost. The aim of the composting is to produce a material for agronomic use stating from source separate collected organic material (e.g. OFMSW). The composting process became very important in the last 20 years in the MSW management and new rules will push up in the future the use of this biological treatment. In particular, the European Council Directive on waste landfill (European Union, Landfill Directive 1999/31/EC) requires member states to draw up strategies to reduce progressively the amount of biodegradable MSW in landfill over a 15-year period to only 35% of the total amount of that produced in 1995. Thus, compliance to this Directive as well as owing to fact that new landfill sites are hardly available because of both shortage of utilizable land and the objections from the residents near the proposed sites will greatly divert the usage of organic fraction of municipal solid waste (OFMSW) for other purposes.

The high moisture content and low calorific power of OFMSW do not allow energy recovery by its direct combustion. On the other hand, as the OFMSW waste is a biodegradable component, the alternative managements to the landfill or incineration are biological processes, such as composting. So far, composting represented the main method that uses OFMSW, and in the last 10 years, composting has been used increasingly. Typically composting is divided in two stages: i. high rate composting during which mostly degradation of easily degradable OM takes place and ii. curing phase during which degradation process slow down and rearrangement of organic molecules occurs (Figure 4).

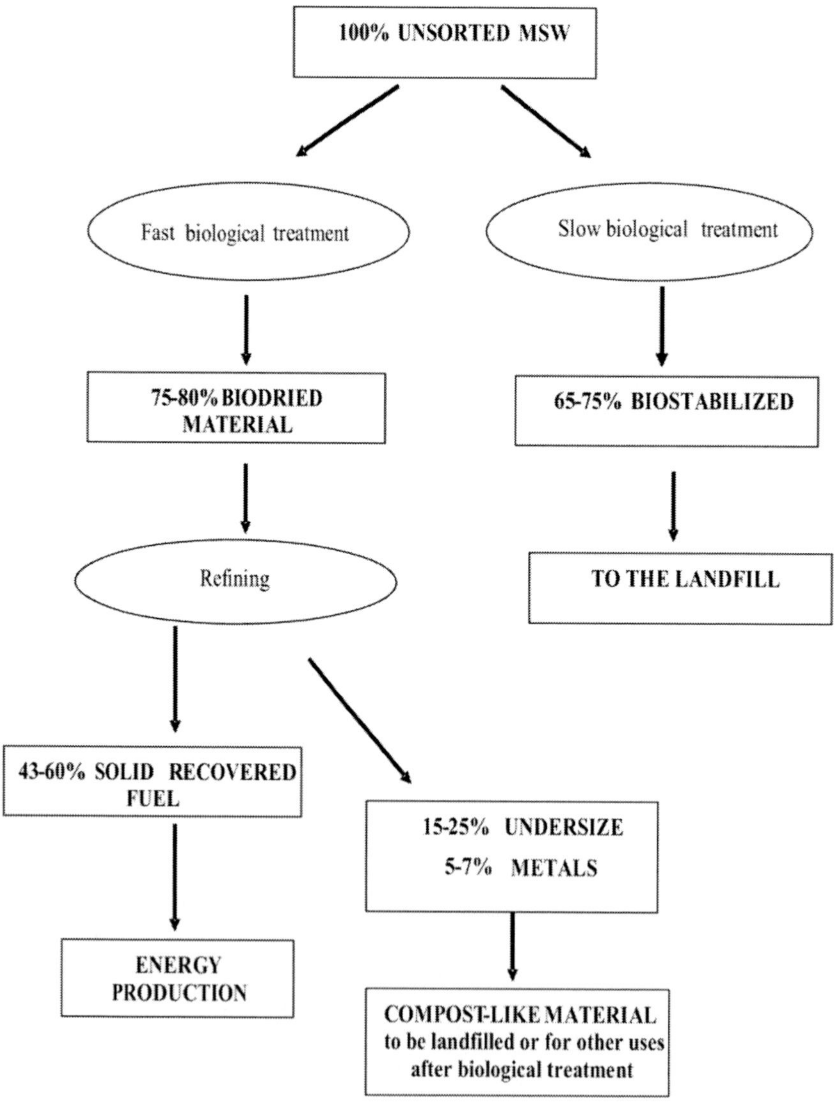

Figure 3. MBTUF mass balance scheme.

3.4.2. The Effect of Aerobic Solid-State Biological Processes on the Reduction of Ghgs Emission

MBT versus Methane Reduction

The biostabilization processes are effectiveness in the waste management to reduce the landfill potential impacts. The organic matter degradation that occurs during biological treatment determines an elevated reduction of biogas production getting low CH_4 emission in landfill (Adani et al., 2002; Forte et al., 2003) (Table 9).

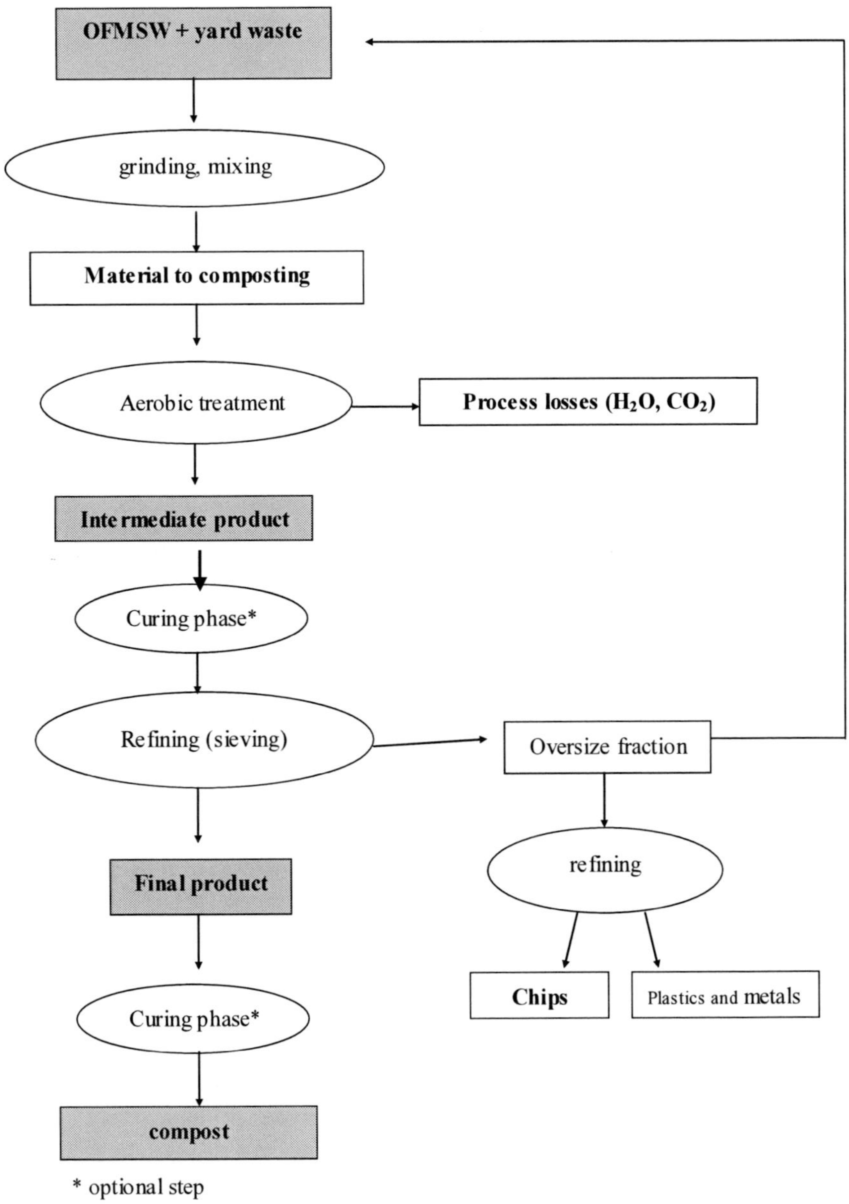

Figure 4. Composting process scheme.

Table 9. Potential waste impact measured for untreated and treated waste

Treatment period		Biogas production tests ($Nl\ kg^{-1}$ dm)	
Test length	21 days	90 days	240 days
0†	-	34	269
5 weeks†	-	90	162
9 weeks†	-	141	153
20 weeks†	-	16	19
Reduction (%)		*52.9*	*92.9*
0‡	347	-	-
20 weeks‡	8	-	-
Reduction (%)	*97.7*		
0§	209.9	-	-
21 weeks§	135.6	-	-
Reduction (%)	*35.4*		
0 weeks #	405	-	-
21 weeks#	24	-	-
Reduction (%)	*94.1*		

† Zach et al., 2000 (MBTSF process).
‡ von Felde and Doedens, 1997 (MBTUF process).
§ Lornage et al., 2007 (MBTUF process).
# Ponsá et al., 2002 (MBTUF process).

Several analytical methods able to measure the anaerobic biogasification potential (ABP) of MSW have been proposed in the past (Binner and Zach, 1999). Biogas laboratory tests were classified in fermentative and incubation tests (Adani et al., 2001). Incubation tests are preferred because they are carrying out under conditions that avoid toxic phenomena (Adani et al. 2001; Scaglia and Adani, 2008a).

The experimental data showed that the MBT process reduce the ABP production at a great extent.

The biogas productions of untreated MSW and OFMSW, and corresponding biological treated products, showed a reduction of biogas of 55.66% and 64.93% after short period of treatment. Considering the ABP composition of 50-60% of CH_4 and 40% of CO_2, the total CH_4 reduction was of 55-65 % (Table 10).

Independently by the kind of test used, biogas detection takes long time to be performed ranging from 21 to 100 days (Binner and Zach 1999; Adani et al., 2001; Hansen et al., 2004) and so, it is not practicable as routine analysis.

To avoid long time consuming analyses, other more quick approaches has been proposed to detect biological reactivity of MSW (Scaglia and Adani, 2008a). Among them, the measurement of respirometric activity had wide application (Cossu and Raga, 2008; Ponsá et al.; 2008; Scaglia et al., 2009). Although respirometric tests refers to the degradation of readily organic matter in short time (1-4 days) they are able to describe the organic matter availability in a longer time (Schievano et al., 2008). So doing several relations were found among respiration indexes and biogas production (Cossu and Raga, 2008; Ponsá et al.; 2008; Scaglia et al., 2009). The DiProVe approach consist in the use of apply the Dynamic Respiration Index (DRI) as indicator of the ABP production.

Table 10. Anaerobic biogasification potential (ABP) and CH4 production measured on MSW and derived products after 4 weeks of treatment (DiProVe data)

Waste typology	ABP ($l\ kg\ dm^{-1}$)	CH_4 † ($l\ kg\ dm^{-1}$)
MSW‡	203±58 n=12	101-122
Biostabilized MSW‡	90±15 n=10	45-54
Reduction (%)	55.66	
OFMSW§	308±71 n=12	154-185
Biostabilized OFMSW§	108±24 n=10	54-65
Reduction (%)	64.93	

† CH_4 =50-60% of ABP.
‡MBTFS processes.
§MBTUF processes.

Scaglia et al. (2009), assuming ABP as dependent variable, significant linear regression with the DRI parameter (R^2=0.89; P< 0.01):

$$ABP= 34.4^{+}0.109\ DRI \quad (2)$$

In Table 11 some unpublished data regarding the measurement of the ABP and of the reduction of the total biogas production measured by DRI, are reported.

Table 11. Dynamic respiration index (DRI), and ABP , determined for MSW and corresponding biological treated products:DiProVe data

Waste typology	Processes number	DRI ($mgO_2\ kg\ dm^{-1}h^{-1}$)	ABP† ($Nl\ kg\ dm^{-1}$)	CH_4 ‡ ($Nl\ kg\ dm^{-1}$)
MSW§	6	1,649±265	214	107-128
Biostabilized MSW§		549±86	94	47-56
Reduction (%)		66.71	55.66	
OFMSW#	4	1,675±1,106	217	108-130
Biostabilized OFMSW#		437±158	82	41-49
Reduction (%)		73.91	62.62	

† ABP= $34.4^{+}0.109$*DRI.
‡ CH_4=50-60 % ABP.
§ MBTFS processes.
# MBTUF processes.

Table 12. Biogas reduction after one year of observation of biological treated waste allocated in a simulated landfill

	Time test (days)	Nm^3/Mg	Biogas Reduction (%)[#]
MSW[†]	2500	51	-
MSW pre-treated 1[†]		24	52.94[#]
MSW pre-treated 2[†]		13	80.39[#]
MSW pre-treated 3[†]		8	84.31[#]
MSW pre-treated 4[†]		5	90.19[#]
MSW pre-treated 5[†]		4	92.15[#]
MSW pre-treated 6 [‡]	500	10	77
MSW pre-treated 7 [§]			

[†] Bockreis and Steinberg, 2005.
[‡] Raninger et al., 1999.
[§] Sormunen et al., 2008.
[#] biogas production percent redaction between MSW and MSW pretreated.

Processes studied refer to both MBTSF and MBTUF processes performed in full-scale plants for 2-4 weeks. Data reported indicated an effective reduction of the potential ABP measured confirming the ABP values measured (Table 11).

Results indicated that mechanical biological treatments are effective in the reduction of ABP as the previous paragraph showed.

Many studies were performed on laboratory-scale (lysimeters) in order to measure the biogas of pre-treated or untreated MSW, under different landfill simulating conditions.

Pre-treated wastes showed a strong biogas reduction (53% to 92%) with respect to untreated MSW (Table 12) (Bidlingmaier et al., 1998; Raninger et al., 1999; Bockereis and Steinberg, 2005). Generally, the results obtained established, also, for pre-treated MSW, a rapid methanogenic phase (Raninger et al., 1999). This because the aerobic pre-treatment degraded the readily degradable OM reducing the production of the volatile fatty acid (VFA) that inhibits the methane production (Vavilin et al., 2006; Sormunen et al., 2008).

Composting *Vs* GHGs Reduction

The MSW contains a large fraction of organic materials which amount depends by MSW origin (Table 1). The production of compost for agronomic use, required to start the process with high OFMSW purity that can be obtained by MSW separated collection.

The use of the OFMSW to produce compost allows the reduction the amount of OM disposed in landfill reducing directly the CH_4 generation.

By using data of biological stability (DRI), in Table 13 are reported the potential reduction of the methane emission from landfill obtained by the diversion of the OFMSW to composting. From our data it can be calculated a reduction, as average, of 76%.

Table 13. Dynamic respiration index (DRI) and anaerobic biogasification potential (ABP), determined for OFMSW and corresponding biological treated products: DiProVe data

Sample classes	VS ($g\ kg^{-1}$ dm)	DRI ($mgO_2\ kg\ dm^{-1}h^{-1}$)	ABP[†] ($Nl\ kg\ dm^{-1}$)	CH_4[†]
OFMSW from separated collection n=22	619.7±120.0	2215±853	276	138-166
compost n=22	485.2±123.8	294±292	67	33-40
Reduction (%)		*86.7*	*75.7*	

[†] ABP= 34.4⁺0.109*DRI.
[‡] CH_4=50-60 % ABP.

The emission of N_2O and CH_4 from the compost is between 1–1,464 mg N_2O m^{-2} day^{-1} and 0–119,000 mg CH_4 m^{-2} day^{-1} depending on the age of compost and the dimension of heaps (Beck-Friis et al., 1999). Other works estimated an emission of 27 g of N_2O Mg^{-1} fresh matter and 293 g of CH_4 Mg^{-1} fresh matter, in nine weeks of composting (Amlinger and Peyr, 2008). Moreover, in the first two months of composting about 30% of organic was decomposed to CO_2 (Ciavatta et al., 1993). No more differences in emission of GHG were produced in MBT, between 110 to 1200 of CH_4 Mg^{-1} fresh matter and 50-200 g N_2O Mg^{-1} fresh matter are been registered in relation to various periods (Amlinger and Peyr, 2008). The production of greenhouse gases is variable and depends very closely by the technical process used.

3.4.3. The Use of Biostabilized Compost-Like Material and Compost to Create C-Sink

MBT versus Carbon Sinks

The tendency of landfill to act as C-sink or C-source depends by the rate of decomposition of MSW in landfill that depends by the biodegradability *vs* recalcitrance of OM. The OM disposed in landfill produce methane because of anaerobic condition.

The existence of unfavorable environmental conditions for the microbial activity (e.g. low pH, toxic substances etc.) determines an apparent recalcitrance of the OM. The literature reported, as typical, inhibition for microbial activity in landfill due to the presence of high volatile fatty acids concentrations and/or low moisture content. In any case, the environmental inhibitions determine, only, a retardation of the biogas production, taking landfill longest time to emit methane.

Bäumler and Kögel-Knaber (2008) studying four different German landfills showed that low moisture content (240 g kg^{-1} wet weight) determined a low degree of microbial transformation and mineralization in the OM after 25 years. Similarly, Kelly et al. (2006) showed no significant degradations of OM because of insufficient moisture content in several landfills. A new approach in landfill management (i.e. bioreactor landfill) allows moistening the MSW in landfill to optimal condition (500-600 g kg^{-1} wet weight) enhancing long-term OM stabilization process (Kelly et al., 2006).

Avoiding any kinds of inhibition, allows methane to be produced in short time optimizing its use as energy source, and lowering the direct emission of CH_4 in the atmosphere.

In presence of optimal environmental conditions, the methane emission depends by the quantity/quality of the OM of MSW landfilled. Some authors consider the total amount of degradable organic matter content (degradable organic fraction) as source to produce methane (Bogner and Etthews, 2003). Nevertheless, different OM fractions show different biodegradable rates under anaerobic conditions. Carbohydrates (cellulosic and no-cellulosic compounds) are the most biodegradable compounds (Barlaz, 2006). Nevertheless, the potential biodegradability of cellulose (C) is partially diminished by the chemical and physical inhibition operated by the associated lignin (L) molecules (Barlaz, 2006; Komilis et al., 2006; Ximenes et al., 2008).

The recalcitrance of lignocelluloses complex under anaerobic condition is know to be one of the most significant factors that causes the accumulations of carbon in the anoxic environmental compartments (sediment, peat, etc). Some authors consider the lignin very recalcitrant because of the incapability of microorganisms to degrade the whole lignin molecules, since initial fragmentations of lignin required oxygen (Komilis and Ham, 2003). In any case, if lignin is partially degraded, the anaerobic degradations occur slower than in aerobic conditions (3 times more slowly) (Benner et al., 1984). On the other hand, some Authors showed the capability of microorganisms to degrade lignin under anaerobic condition (Benner et al., 1984). It can be suggested that the cellulose/lignin ratio (C/L) is an efficacy indicators of the recalcitrance degree of OM in landfill (Barlaz, 2006; Komilis et al., 2006; Ximenes et al., 2008).

Kelly et al. (2006) showed for a MSW landfill the degradation of cellulose, the concentration of lignin-like complex and the decrease of the C/L ratio (Table 14).

The trend of C/L during the landfill life showed a progressive diminution reaching low values: an exponential model was set up ($C/L=1.73*e^{-0.0682*year}$, $R^2=0.42$, $P<0.001$). Values of C/L less than 0.5 - 0.2 are considered the limit below these, the waste can be considered stable (Barlaz, 2006; Kelly et al., 2006).

Table 14. Relative and absolute cellulose and lignin degradation in a landfill (Riverband landfill) after 11 years (Kelly et al., 2006)

Time	VS	Cellulose (C)	Lignin (L)	C/L
year	g kg^{-1} dm	g kg^{-1} dm		
		g		
2	476.2±11.2	115.6±66	141.6±24	0.82
n=2		*115.6*†	*141.6*†	
5	308.5±123.5	61.7±50.8	131.4±52.3	0.47
n=4		*48.7*†	*103.7*†	
11	58.1±145.3	42.2±23.0	184.9±112.5	0.23
n=4		*23.4*†	*102.8*†	
Reduction (%)	-	*79.75*	*27.40*	*71.95*

† Absolute content.

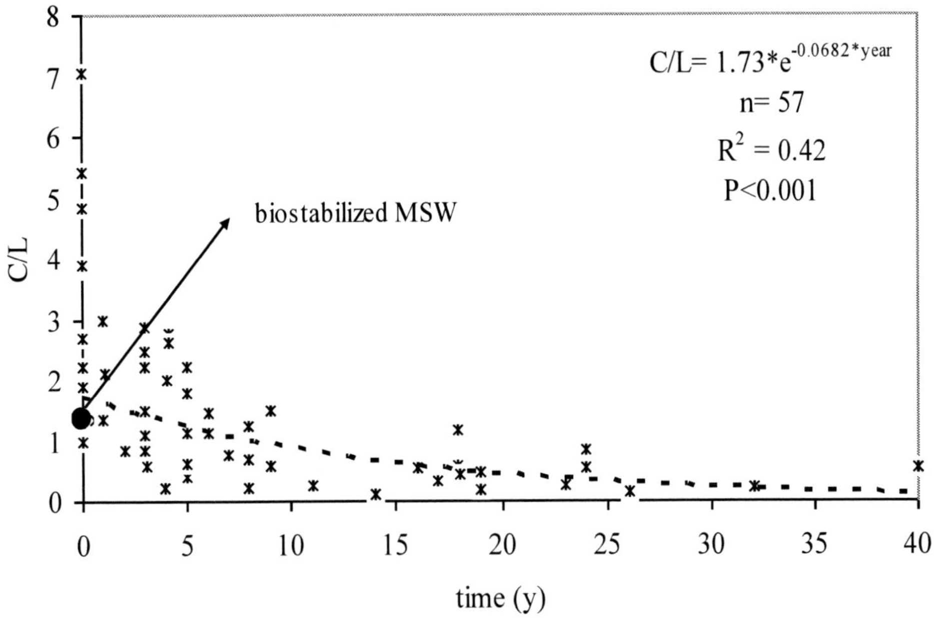

Figure 5. Trend of the C/L ratio of OM of MSW in landfill during the time (experimental data from Barlaz, (2006), Kelly et al. (2006), Komilis et al. (2006), DiProVe elaboration).

The application of the exponential model suggest that at C/L of 0.2-0.5 correspond to a landfill age of 18-32 years (Figure 5).

The recalcitrance of OM of MSW has been evaluate correlating C/L ratio and methane production (BMP) for MSW landfilled for 11 years (Kelly et al., 2006).

A good linear regression was found: BMP=114.8*C/L-31.1 (n=12, R^2=0.85, P<0.05).

A BMP below 26 Nl kg $^{-1}$ dm correspond to C/L of 0.2-0.5.

Previous chapter have shown MBT process to be able to degrade labile organic fraction preserving more recalcitrant ones. Macromolecular composition indicated that modification occurred during aerobic process lead to an end product rich in ligno-like complex (Table 7) responsible for the low degradability of the biostabilized fraction such as biological methods assessed (Table 6). The C/L calculated from MBT processes (Table 8) as been compared with those from literature for landfilled MSW (Figure 4) and corresponding landfilling time calculated. Interesting was the fact that these MBT treated MSW, showed C/L ratios similar to those of MSW landfilled for more than 5 years (Figure 5).

No data exist about biogas produced by MBT treated-MSW in landfill. Anyway, the comparison of data relative to potential methane production (BMP) of landfilled MSW, with those of biostabilized MSW, allows getting some indication of the effect of MBT on the recalcitrance of OM (Table 15).

Data reported in Table 15 are very interesting because they allow estimating the amount of MSW carbon that can be accounted as credit in the GHGs balance.

In agreement with literature, data obtained suggest that MSW characterised by MBP below 20-25 Nl kg^{-1} dm can be considered as recalcitrant. That means that about 200 kg per Mg of MSW dry matter (dm) (Table 15) can be stocked in landfill, that is about 40-45 % of the total MSW carbon (assuming starting MSW-C content of 30% - data DiProVe).

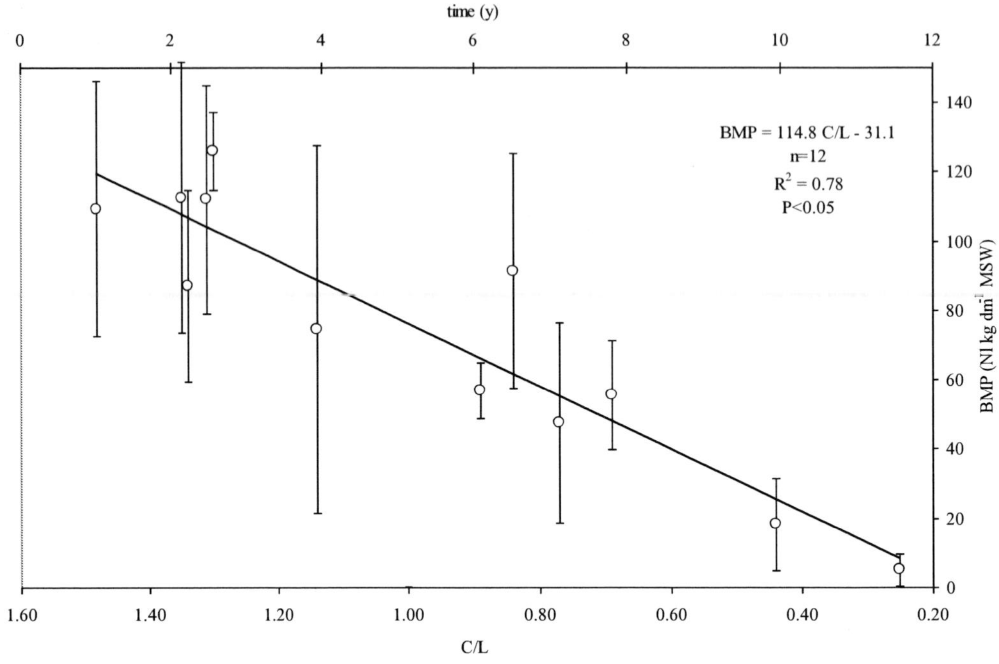

Figure 6. Trend of the C/L ratio and corresponding methane potential production (BMP) of MSW in landfill during the time (data from: Kelly et al. 2006, DiProVe elaboration).

Table 15. Comparation of values of C/L, BMP and DRI of MSW biostabilized and MSW landfilled

Treatment	C/L† (w/w)	BMP ($Nl\ kg^{-1}$ dm MSW) †	DRI ($mgO_2\ kg\ dm^{-1}\ h^{-1}$) †	C ($g\ kg^{-1}$ dm) †	Corresponding age in landfill (years) ‡	Corresponding age in landfill (years) §
MBT (after 4 weeks of treatment)	1.2-1.4	60-80	500-600	220-250	3-5	3.5-5
MBT as before plus curing phase	0.35-0.45	10-20	100-150	180-200	20-23	10-11

† DiProVe data.

‡ Data calculated by considering C/L ratio of this table and the equation reported in Figure 5.

§ Data calculated by considering BMP of this table and the equation reported in the text (BMP=114.8*C/L-31.1).

Assuming an MBT process of 4-5 weeks, as typical in Italy, giving a MSW characterised by a MBP of 60-80 Nl kg^{-1} dm and DRI of 500-600 $mgO_2\ kg\ dm^{-1}\ h^{-1}$, and performing a complete C-mass balance, it can be calculated that starting 100 kg of MSW-C coming from undersize fraction (size < 60 mm = organic fraction MSW), 40 kg are transformed in CO_2 by aerobic biological process; 17.8 kg are transformed in biogas (60% CH_4, 40% CO_2) and 41.8 kg are preserved as recalcitrant carbon in landfill.

Therefore in term of GHGs balance MBT process allows to avoid the production of 24 kg of CH_4 and store 41.8 kg of recalcitrant C in landfill.

Composting Versus Carbon Sink

With increasing CO_2 emissions into the atmosphere and the possible consequences for global climate change, there has been a growing interest in studying the contribution of sequestering carbon (C) in terrestrial ecosystems, i.e. the soil (Wang et al., 2004). In fact, although CO_2 is primarily produced by fossil fuel combustion, land use changes have contributed significantly to atmospheric CO_2 increases through the conversion of forest and grassland to agricultural land (Mosier, 1998).

The soil organic carbon (SOC) pool is the largest terrestrial reservoir of carbon (Lal, 2001). In fact, global soil carbon is estimated to be 1200-1600 Pg and the estimated historic losses of C from agricultural soils amounts to 40-60 Pg C (Paustian et al., 1997). These same authors argue that the above losses could represent a reference level for C sequestration potential, i.e. the restoration to the native C level represents the potential for C sequestration. In its Second Assessment Reports The Intergovernmental Panel on Climate Change estimated that it may be possible to sequester 40 to 80 Pg of C in cropland soils in the next 50-100 years, thereby confirming the above data (Cole et al., 1996). Thus, as agricultural soils are relatively C-depleted, they represent a potential CO_2-sink if lost C can be regained (Paustian et al., 1997; Vleeshouwers and Veraghen, 2002).

Sequestering carbon in agricultural soil means decreasing the output and increasing the input (Izaurralde et al., 2001). Soil C content can be increased through correct soil management: high residue production, mulching, tillage, reduction of bare fallow and fertilization and land use change (Batjes, 1998).

The agronomic utilization of organic wastes such as livestock effluents, sewage sludge, and compost has been increased due to the need of lowering disposal costs, recycling nutrient elements in the soil crop system and, to counter the reduction of soil organic matter content (Giusquiani et al., 1995). Consequently, waste biomasses can be successful employed to increase cropland soil C content (Lal, 2001). Smith et al. (2005) reported that organic farming is the only C enhancing practice that has increased from 1990 to 2000 and that this practice is set to further increase between 2000 and 2010. However, the same authors argue that the use of organic sources in one place reduces the organic inputs in another, resulting in a zero net soil-C gain. Nevertheless, the European Council Directive on waste landfill (Landfill Directive 1999/31/EC) requires member states to draw up strategies progressively reduce the amount of biodegradable municipal solid waste in landfill over a 15-year period to only 35% of the total amount of that produced in 1995. Compliance to this Directive will greatly increase the amount of (organic matter) OM to be composted and used in agriculture. For example, in Italy the production of high quality compost has increased from approximately 200,000 Mg in 1993 to 2,700,000 Mg in 2004 and is predicted to further double in the next 10 years, with similar data being reported in other EC countries (Landfill Directive 1999/31/EC).

Little data exists on the effects of using compost on C sequestration. Smith et al. (2005), discussing the potential of C sequestration in Europe, underlined the lack of data on compost management as a means of C sequestration. The fact that generally studies dealing with compost use are focused on the agronomic and chemical effects of compost on soil rather that on C sequestration, explains the lack of available data (e.g. Gonzàles-Vila et al., 1999; Quèdraogo et al., 2001; Leifeld et al., 2002; Adani et al., 2006).

The term compost, indicates a material from various origins and with a wide range of compositions that has undergone composting (Richard and Woodbury, 1992). Compost is a humic-like material, which acts as an organic amendment due to the high organic matter content. The composting process proceeds through the degradation of the labile organic fraction and the concentration of more recalcitrant fractions, resulting in biomass enrichment of the humified materials (Lhadi et al., 2006). Compost products vary depending on the raw material and composting technology used (Richard and Woodbury, 1992). The organic matter quality plays a primary role (Paustian et al., 1997) between the different parameters that affect OM decomposition in soil (e.g. C/N ration, soil pH, nutrients availability, bulk density etc.).

The OM quality is a key concept in soil organic matter studies and can be defined as the capacity of OM to be utilized by soil microbes as a source of energy (Rovira and Vallejio, 2002). Many parameters and indexes have been proposed in the past to measure the OM quality, e.g. C/N, lignin/N and cellulose/lignin/N ratios (Rovira and Vallejio, 2002). The OM composition with regard to the macromolecular components appears to be more important than other parameters. For example, lignin, phenols and tannin contents in the OM, were more important in determining the soil OM decomposition than the C/N ratio (Paustian et al., 1997). The more widely used models describing the soil-C cycle (e.g. CENTURY model), divided soil-C and litter in different pools in relation to their degradability (quality) (Izaurrulde et al., 2001). As a strong correlation exists between the chemical and biological stability of OM, the ratio of chemical-labile versus chemical-recalcitrant compounds is generally used to describe OM quality (Six et al., 2002; Mikutta et al., 2006). Rovira and Vallejio (2000) described plant residue degradability in soil by dividing the OM in a labile and a recalcitrant pool, depending on the solubility of the OM in strong acids. These author determined that the recalcitrant fraction content ranged from 250 g kg^{-1} to 650 g kg^{-1} , in the order Medicago<Quercus<Eucalyptus<Pinus. Adani et al., (1995) have divided the labile and recalcitrant OM fractions of compost by using a procedure similar to the Klason lignin isolation. The partition of compost OM in recalcitrant and labile pools is very useful, as the former is preserved and the latter is degraded after soil incubation (Rovira and Vallejio, 2002; Adani and Ricca, 2004a; Adani and Spagnol, 2006). A simple representation of the recalcitrance of soil OM is reported in Table 16.

The total amount of carbon stored in soil result from the net balance between input and output. Complete mass balance (carbon balance), before and after the soil amendment, enabled tracing of the sequestered C (Adani et al., 2006). An alternative to monitoring the disappearance or the sequestering of C in soil is the measurement of the rate at which CO_2 is produced during amendment decomposition (Rochette et al., 1999). This approach enables the measurement of instantaneous decomposition rates and provides information on the same time scale as changes of the variables that modulate decomposition of residues in soils (e.g. soil temperature, moisture and C availability) (Rochette et al., 1999).

Table 16. Organic matter recalcitrance in soil (based on Kögel-Knabner et al., 2008 and Lützow et al., 2006)

Selective preservation	Primary recalcitrance: The indigenous molecular characteristic of plant tissue, constitutes by aromatic and alkyl compound (lignin, cutin, suberin, tanninis), that are resistant to biodegradation. Secondary recalcitrance: Microbial and faunal-derived compounds, such as: microbial residues rich in murein and chitin; pigments with aromatic structure and chitinous cuticles are often synthesized by soil animals.
Spatial inaccessibility	Occlusion of organic matter in soil aggregation that makes inaccessible to decomposer organism.
Interaction with mineral and metal	Creation of linkages, between organic matter with minerals and metal ions, resistant to microbial and enzymatic deconstruction.

Table 17. Organic matter sequestration by compost application

Reference	SOC[†] control g kg^{-1} dm[§]	TOC[‡] compost g kg^{-1} dm[§]	Compost application Mg ha^{-1} [§]	Increase of TOC %	Time after application
Adani et al., 2008	18.75	198.9	34	3.53	150 days
			56	10.74[#]	
Weber et al., 2007	5.35	137	36	17.19	1 year
			72	42.24[#]	
		166	36	20.56[#]	
			72	62.99[#]	
	5.83	137	36	19.69[#]	1 year
			72	31.16[#]	
		166	36	14.04	
			72	34.25[#]	
	5.09	137	36	25.34[#]	1 year
			72	32.61[#]	
		166	36	25.15[#]	
			72	38.31[#]	
Bastida et al., 2008	15	406	120[π]	83[#]	19 months

[†] SOC = starting soil organic carbon.

[‡] TOC = compost total organic carbon.

[§] Dry matter.

[#] Significant differences at the level $P<0.05$.

[π] Fresh matter.

Estimating a precise life-time for soil organic matter derived from compost addition is very difficult (Favoino and Hogg, 2008). The turnover time of various fractions of soil organic carbon is different. For example it was estimated a life-time less than one year for microbial biomass and a range between 5 and 1,000 years for organic matter associated with soil minerals (Favoino and Hogg, 2008). Applying first-order decay kinetics it has been estimated that only 2 % of carbon applied to the soil by compost, would remain in the SOM after 100 years (Favoino and Hogg, 2008).

Delschen and Vetterlein (1997) found accumulation rates between 0.02 and 0.08% SOM per years, depending on treatment and application rate, and the values decreased with time. On grasslands, Riem-Vis (1984) found an increase of 0.2% per 25 m3 compost given annually (Zwart, 2003).

Recent field experiments on the application of compost to the soil, has highlighted that a contributions of 34-72 Mg ha^{-1} of compost was able to increase, in the short time, the content of carbon in the soil of 3.5-62.99 % (Table 17).

A model to predict a long-term potential C-sequestration in soil amended with compost, was reported by Hogg (2002). Model proposed that compost applied, partially is stabilized becoming part of the stable SOM. This fraction could be mineralized but at a low degradation rate (Figure 7).

The scheme reported in Figure 7, suggests that the rate of accumulation of compost-OM in the SOM is higher than those mineralized. On the other hand the stable SOM is mineralized at a rate lower then that of the of compost-OM accumulation. Therefore the net result is a conservation of compost OM.

Anyway assuming the model parameters as in following: a = 3.5%, x =20%, z=1.5 %, y = 10% (Nortcliff and Amlinger 2003) it can be calculated the effects of different rate of compost application on the level of organic matter in the soil (Figure 8).

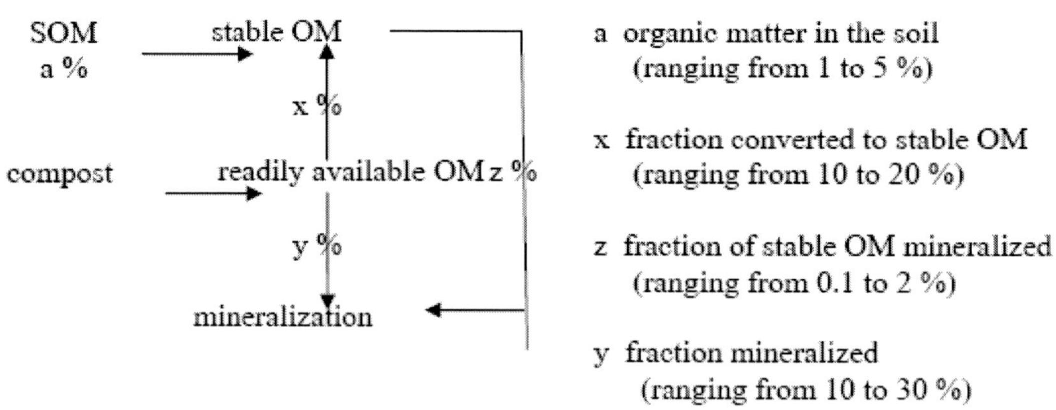

Figure 7. Model proposed by Hogg et al. 2002, Nortcliff and Amlinger, 2003.

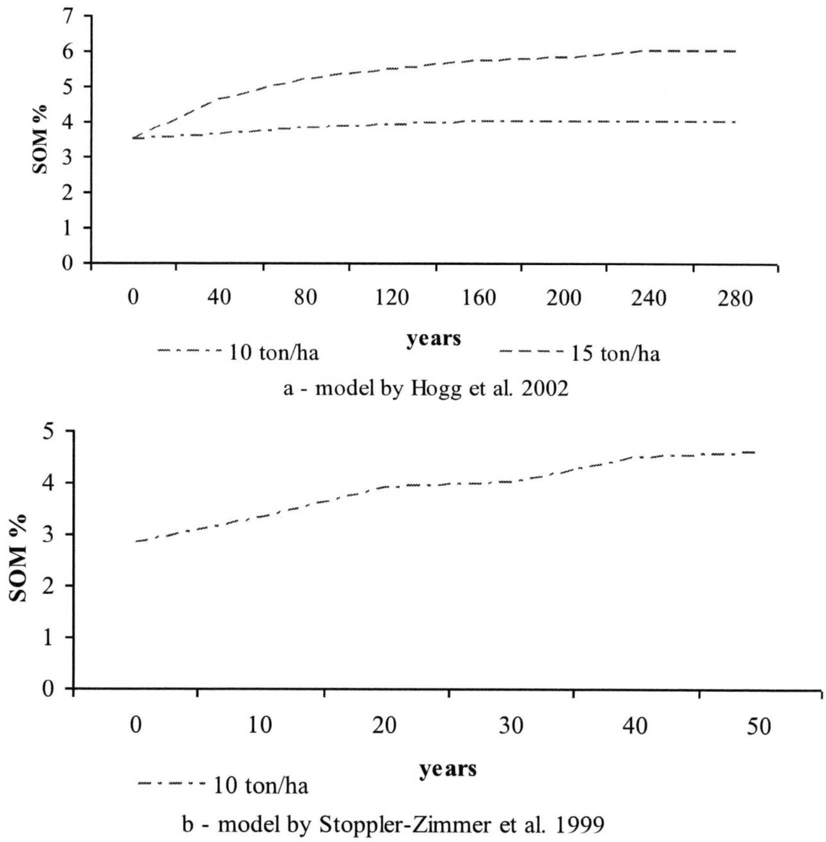

Figure 8. Models to predict long-term application of compost.

Computer simulation model (Stoppler-Zimmer et al., 1999) was developed to predict the long-term compost management in loamy soil (Figure 8). Model suggested an increasing of SOM during first years of application that could be kept stable later. Highest compost allowed SOM to increase continually over 30-40 years.

This approach is very simple and do not consider many and more complicated aspects related to the OM evolution in the soil, above all relate to the mechanism of C sequestration in soil that depends by soil characteristics. For example, Table 17 reported, for compost dose higher than those considered in the model (34 Mg Ha^{-1}), no net compost-OM accumulation, indicating a degradation rate much higher than that reported in the model proposed.

Long term experiments, better than model, could reveal the effect of compost on SOM (Zwart, 2003).

Field long-term experiments performed by Zwart (2003) and De Hann and Lubbens (1984) with MSW compost, suggested that the SOM, fundamentally, remained constant or little increased (Figure 9).

Therefore more data needs to understand the exact role of compost from MSW to increase SOM and therefore to understand if soil can act as C-sink. Surely, compost avoiding SOM decrease help to keep constant the total C stored in agricultural soil.

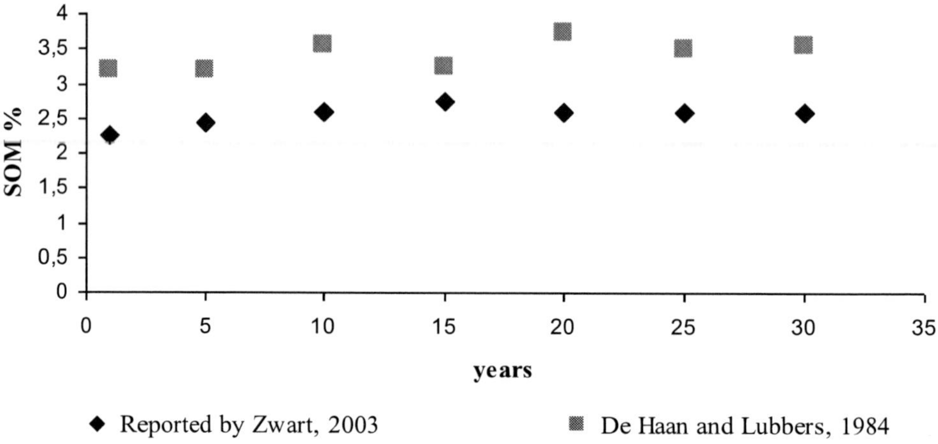

Figure 9. Long-term application of compost.

References

Adani, F.; Genevini, P.L.; Tambone, F. *Compost Sci Util.* 1995, 3, 25-37.

Adani, F.; Genevini, P.L.; Gasperi, F.; Zorzi, G. *Compost Sci Util.* 1997, 5, 53-62.

Adani, F.; Tambone, F.; Genevini, P.L.; Calcaterra, E. *Techno Enviro.* 1998, 4, 2-8.

Adani, F.; Scatigna, L.; Genevini, P.L. *Waste Manage Res.* 2000, 18, 1-9.

Adani, F.; Baido, D.; Calcaterra, E.; Genevini, P.L. *Bioresource Technol.* 2002, 83 (3), 173-179.

Adani, F.; Confalonieri, R.; Tambone, F. *J Enviro Qual.* 2004, 33, 1866–1876.

Adani, F.; Spagnol, M.; Genevini, P. *Biogeochemistry* 2006, 78, 85-96.

Adani F.; Spagnol M.; Nierop, K.G.J. *Biogeochemistry* 2007, 82, 55-65.

Adani, F.; Tambone, F.; Genevini, P. *Waste Manage.* 2009, 29, 174-179.

Amlinger, F.; Peyr, S., *Waste Manage. Res.* 2008, 26, 47-60.

Asomani-Boateng, R.; Haight, M. *Third World Plan Rev.* 1999, 21 (4), 411- 428.

Aronica, S.; Bonanno, A.; Piazza, V.; Pignato, L.; Trapani, S. *Waste Manage.* 2009, 29, 233-239.

Barlaz, M.A. *Waste Manage.* 2006, 26, 321-333.

Barrena Gómez, R.; Vázquez Lima, F.; Sánchez Ferrer, A. *Waste Manage Res.* 2006, 24, 37-47.

Bastida, F.; Kandeler, E.; Moreno, J.L., Ros, M.; Garcìa, C.; Hernàndez, T. *Appli. Soil Ecol.* 2008, 40, 318-329

Bäumler, R.; Kögel-Knaber, I. *J. Environ. Qual.*2008, 37, 146-153.

Beck-friis, B.; Pell, M.; Sonesson, U.; Jönsson, H.; Kirchmann, H.. *Environ. Monit. Assess.* 2000, 62, 317-331.

Batjes, N. H. *Eur.J. Soil Sci.* 1996,. 47, 151–163.

Benner, R.; Maccubbin, .E.; Hodson, R.E. *App. Environ. Microbiol.,* 1984, 47 (5), 998-1004.

Bidlingmaier, W.; Sidaine, J.; Papadimitriou, E.K. *Rev Environ Sci Bio/Techno.* 2004, 3, 307–320.

Binner, E.; Zach, A.; *Waste Manage Res.* 1999, 17, 543-554.

Bockreis, A.; Steinberg, I. *Waste Manage.* 2005, 25, 337-343.

Bogner, J.; Spokas, K. *Chemosphere.* 1993, 26, 366– 386.

Bogner, J.; Metthews, E. *Global biochemical cy.* 2003, 17 (2), 1065.

Borjesson, G.; Ph.D. thesis. Dep. of Microbiol., Swedish Univ. of Agric. Sci., Uppsala, SE, 1996.

Bozkurt, S.; Moreno, L.; Neretnieks, I. *Sci Total Environ.* 1999, 228: 135- 152.

Brown, S.; Kruger, C.; Subler, S. *J. Environ. Qual.* 2008, 37, 1396-1410.

Calcaterra, E.; Baldi, M.; Adani, F. 2000. *An innovative technology for municipal solid waste energy recovery.* C.I.P.A.-Centro di Ingegneria per la Protezione dell'Ambiente: Padoa, IT, 2000;123-135.

CEN/TC 343/WG 3. *Solid recovered fuels-Determination of current rate of aerobic microbial activity using the real dynamic respiration index.* CEN/TC 343 N159 stage 32, 2005.

Chesson, A. In *Driven by nature Plant Litter Quality and Decomposition.* Cadisch, G.; Giller, K.E. Eds.; CAB International: Wallinford, UK 1997, pp 47-66.

Ciavatta, C.; Govi, M.; Pasotti, L.; Sequi, P. *Bioresour. Technol.* 1993, 43, 141.

Cole, C.; Cerri, C.; Minami, K.; Mosier, A.; Rosemberg, N.J.; Sauerbeck, D. In: *Climate Change 1995: Impacts, Adaptation and Mitigation of Climate Change,* Watso RT, Zinowera MT, Moss RH, Eds. 1996, pp. 745-771. Cambridge University Press. Cambridge.

Cossu, R.; Raga, R. *Waste Manage Res.* 2008, 28, 381-388.

Czepiel, P.; Mosher, B.; Harriss, B.; Shorter, J.H.; McManus, J.B.; Kolb, C.E.; Allwine, E.; Lamb, B.J. *Geophys Res.* 1996, 101(D11), 16,711–16,719.

Derham, J. In *The role of groundwater in sustainable development, proceedings of 15th annual groundwater seminar.* International Association of Hydrogeologists: Portaloise, USA, 1995.

D'Imporzano, G.; Crivelli, F.; Adani, F. *Sci. Total Environ.*, 2008, 402 (2-3), 1 278-284.

Driessen, J.H.A.; Moura, M.L.; Korzilius, E.P.A.; van der Sloot, H.A. 1995. *The Sustainable Landfill.* In: Christensen, T.H. Cossu, R., R. Stengmann, R. Eds. CISA: Cagliari, IT, 1995; pp. 15-24.

El-Fadel, M.; Findikakis, A.N.; Leckie J.O.J.; *Enviro Manage.* 1997, 50, 1–25.

El-Fadel, M.; Bou-Zeida, E.; Chahineb, W.; Alaylic, B. *Waste Manage.* 2002, 22, 269-282.

European Committee For Standardization. *Solid recovered fuels - Method for the determination of biomass content.* European Committee For Standardization: Brussels, BE, 2006. prCEN/TS 15440:2006 (E).

European Union, *Landfill Directive 1999/31/EC.ù*

EUROSTAT. In *Waste Generated and Treated in Europe. Data 1995-2003.* European Commission –Eurostat: Brussels, BE, 2006, 131 p.

Favoino, E., Hogg D. *Waste Manage. Res.* 2008, 26, 61-69.

Forte, A.; Maggi, G.; Scaglia, B.; Adani, F. 2003. In *Ninth International Waste Management and Landfill Symposium;.* In Christensen, T.H. Cossu, R., R. Stengmann, R. Eds. S. Margherita di Pula, IT, 2003.

Fricke, K.; Santen, H.; Wallmann, R. *Waste Manage.* 2005, 25, 799-810.

Galle, B.; Samuelsson, J.; Svensson, B.; Borjesson, G. *Environ Sci Technol.* 2001, 35, 21-25.

Giusquiani, P.L.; Pagliai, M.; Gigliotti, G.; Businelli, D.; Benetti, A. *J. Environ. Qual.*, 1995, 24, 175-182.

Gonzàles-Vila, F.J.; Almendros, G.; Madrid, F. *Sci Total Environ*, 1999, 236, 215-229.

Halvadakis, C.P.; Robertson, A.P.; Leckie, J.O. *Landfill methanogenesis: Literature review and critique.* Stanford Univ. Dep. of Civ. Eng. Tech. Rep. 271, Stanford Univ., Stanford, USA, 1983.

Hamoda, M.F.; Abu Qdais, H.A.; Newham, *J. Res, Conserv Recyc.*, 1998 23, 209-223.

Hansen, T.L.; Schmidt, J.E., Angelidaki, I.; Marca, E.; Jansenb J.; Mosbæk, H.; Christensen T. H. *Waste Manage.* 2004, vol 24, 393–400.

Hedges, J.J.; Keil, R.G. *Mar Chem.* 1995, 49, 81-115.

Hedges, J.J.; Oades, J.M. *Org Geochem.* 1997, 27, 319-361.

Himmel, M.E.; Ding S.Y.; Johnson, D.K.; Adney, W.S.; Nimlos, M.R.; Brady, J.W.; Foust, T.D. *Sci.* 2007, 315, 804-807.

Hogg, D.; Favoino, E.; Nielsen, N.; Thompson, J.; Wood, K., Penschke, A.; Economides, D.; Papageorgiou, S. *Economic analysis of options for managing biodegradable municipal waste.* Final Report to the European Commission. 2002.

Hovde, D.C.; Stanton A.C.; Meyers T.P.; Matt D.R.J. *Atmos Chem.* 1995, 20, 141–162.

Iannotti, D.A.; Pang, T.; Toth, B.L.; Elwell, D.L.; Keener, H.M.; Hoitink, H.A.J. *Compost Sci Util.* 1992, 1 (3), 52–65.

Kelleher, B. P.; Simpson, A. J. *Environ. Sci. Technol.* 2006, 40, 4605-4611.

Kelly, R.J.; Shearer, B.D.; Kim, J.; Goldsmith, C.D.; Hater, G.R.; Novak, J.T. *Waste Manage.* 2006, 26, 1349-1356.

Kjeldsen, P.; Barlaz, M. A.; Rooker, A. P.; Baun, A.; Ledin, A.; Christensen, T. H. *Crit Rev Env Sci Tech.* 2002, 32(4), 297-336.

Kögel-Knaber, I., Leeuw, J.W.; Hatcher, P.G.; Leeuw, J.W. *Sci Total Environ.* 1992, 117 (118), 175-185

Kögel-Knabner, I. *Soil Biol Biochem.* 2002, 34: 139-162.

Kögel-Knaber, I.; Guggenberger, G.; Kleber, M.; Kandeler, E.; Kalbitz, K.; Scheu, S.; Eusterhues, K.; Leinweber, P. *J Plant Nutr Soil Sci.* 2008, 171, 61-82.

Komilis, D.P.; Ham, R.K.; Stengmann, R. *Waste Manage Res.* 1999, 17, 10-19.

Komilis, D.P.; Ham, R.K. *Waste Manage.* 2003, 23, 419-423.

Lal, R. *Adv. Agrono.*, 2001, 71, 145-191.

Lasaridi, K.E.; Stentiford, E.I. *Water Res.* 1998, 32(12), 3717-3723.

Leikam, K.L.; Stegman, R. *La nuova discarica controllata*: Teolo (IT), May 19-21 2004, on CD.

Lelieveld, J.; Crutzen, P.J., Dentener, F.J. *Tellus B.* 1998, 50, 128-150.

Lhadi, E.K.; Tazi, H.; Aylaj, M.; Tambone, F.; Adani, F. *Compost Sci Util.* 2006, 12 (2): 137-144.

Lorenz, K.; Lal, R.; Preston, C.M.; Nierop, K.G.J. *Geod.* 2007,142, 1-10.

Lornage, R.; Redon, E.; Lagier, T.; Hébé, I.; Carrè, J. *Waste Manage.* 2007, 27, 1755-1764.

Morris, J. Ph.D. thesis, Fac. of Eng., Univ. of the Witwatersrand, Johannesburg, ZA, 2001.

Mosher, B.W.; Czepiel, P.; Harriss, R.; Shorter, J.; Kolb, C.; McManus, J. B.; Allwine, E.; Lamb, B. *Environ Sci Technol.* 1999, 33, 288– 294.

Mosier, A.R. *Biol. Fertil. Soil.*, 1998 27, 221-229.

Leifeld, J.; Siebert, S.; Kogel-Knaber, I. Changes in the chemical composition of soil organic matter after application of compost. *Eur. J. Soil Sci.*, 2002, 53, 299-309.

Müller, W.; Fricke, K.; Vogtmann, H. *Compost Sci Util.* 1998, 6 (3), 42–52.

Münnich, K.; Mahler, C.F.; Fricke, K. *Waste Manage.* 2006, 26, 150–157.

Nortcliff, S.; Amlinger, F. In: *Appying compost benefücts and needs*, 2001, Federal Ministry of Agriculture, Forestry, Environment and Water Management, Austria, and European Communities Eds, Brussels 23-23 November 2001, ISBN 3-902 338-26-1, 19- 36, 2003.

Nozhevnikova, A.N.; Lifshitz, A.B.; Lebedev, V.S.; Zavarin, G.A. *Chemosphere.* 1993, 26, 401– 417.

Quèdraogo, E.; Mando, A.; Zombrè, N.P. *Agric. Ecosyst. Environ.*, 2001, 84, 259-266.

Paustian, K.; Andrén, O.; Janzen, H.H. *Soil Use Manage.*, 1997, 13, 230-244.

Pichler, M.; Kögel_Knaber, I. J. Environ. Qual. 2000, 29, 1337-1344.

Ponsá, S.; Gea, T.; Alerm, L.; Cerezo, J.; Sánchez, A. *Waste Manage*, 2008 available online, doi:10.1016/j.wasman.2007.12.002.

Raninger, B.; Nelles, M.; Harant, M.; Steiner, G.; Staber, M.; Lorber, K.E (1999). In *Proceedings Sardinia 97*. Christensen, T.H. Cossu, R., R. Stengmann, R. Eds., Six International Landfill Symposium, S. Margherita di Pula, IT, 1999.

Rasse, D.; Rumpel, C.; Dignac, M.F. *Plant Soil.* 2005, 269, 341-356.

Read, A.D.; Hudgins, M.; Phillips, P. *Waste Manage.* 2001, 21, 617-629.

Revans, A.; Ross, D.; Gregory, B.; Meadows; M.; Harries C.; Gronow, J. (1999). In *Sardinia '99, Seventh International Landfill Symposium.* Christensen, T.H., Cossu, R., Stegmann, R. Eds.; CISA: S. Margherita di Pula, IT, 1999. pp. 199-206.

Richard, T.L.; Woodbury, P.B. *Biomass Bioenergy*, 1992, 3, 195-201.

Robinson, H.D.; Knox, K.; Bone, B.D.; Picken, A. *Waste Manage.* 2005, 25, 383-391.

Rovira, P.; Vallejo, V.R.; *Geoderma.* 2002, 107, 109-141.

Said-Pullicino, D.; Gigliotti, G. *Chemosphere* 2007, 68, 1030–1040.

Scaglia, B.; Adani, F. Sustainable Landfill and Biological Stability. In: Waste Management: Research, Development and Policy. *Ed.* J.R. Lavelle. NovaPublisher. Binding: Hardcover; Pub.Date:2008, 2ndQuarter ISBN:978-1-60456-256-9.

Scaglia, B.; Erriquens, F.L.,; Gigliotti, G.; Taccari, M.; Ciani M.; Genevini, P.L.; Adani, F. *Bioresour. Technol.* 1993, 98, 706-713.

Scaglia, B.; Adani, F. *Total Environ.* 2008, 394, 183-191.

Scaglia, B.; Confalonieri, C.; D'Imporzano, G.; Adani, F. *Bioresour. Technol.* 2009. submitted.

Schievano, A.; Pognani, M.; D'Imporzano, G.; Adani, F. *Bioresour. Technol.* 2008, 99, 8112–8117.

Science and engineering of composting; In: Hoitink, H.A.J.; Keener, H.M. (Eds.), Science and Engineering of Composting: Design Environmental Microbiological and Utilization Aspects. Ohio State University: Columbus, (OH),1993. p. 728.

Smith, P. *Eur J Soil Sci.* 2005, 56, 673–680

Sormunen, K.; Ettala, M.; Rintala, J. *Waste Manage.* 2008, 28 (1), 151-163.

Sparks, D.L. Soil physical chemistry. CRC Press, Boca Raton (USA), 1999.

Stevenson, F.J. Humic chemistry genesis, composition, reactions. 2nd ed. John Wiley and Sons, New York, 1994.

Stöppler-Zimmer, H., Gerke, H. H., Arning, M., 1999,. In: *Proceedings ORBIT 99*, part II, 453-461, 1999.

Surflita, J.M.; Gerba, C.P.; Ham, R.K.; Palmisano, A.C.; Rathje, W.L.; Robinson, J.A. *Environ Sci Technol.* 1992, 26, 1486-1494.

Taylor, R.; Allen, A. In: *Waste disposal and landfill: Information needs. Chapter 12.* 2001. www.crdi.ca/en/ev-33948-201-1-DO_TOPIC.html.

UNI (Italian Organization for Standardization). Potential dynamic respiration index (PDRI) determined using method UNI/TS 11184:2006 (Waste and refuse derived fuels - Determination of biological stability by dynamic respirometric index) UNI: Milan, IT, 2006. (Italian language).

US Composting Council. 1997. In: Leege, P.B., Thompson, W.H. (Eds.), *Test Methods for the Examination of Composting and Compost.* The US Composting Council: Bethesda, USA, 1997.

USEPA. United States Environmental Protection Agency. USEPA-600/9-81-0029. 2003. www.epa.gov/epaoswer/non-hw/muncpl/pubs/msw05rpt.pdf.

Van Soest, P.J.; Robertson, J.B.; Lewis, B.A. *J. Dairy Sci.* 1991, 74, 3583-3597.

Vavilin, V.A.; Jonsson, S., Ejlertsson, J.; Svensson, Bo H. *Biodegradation.* 2006, 17 (5), 389-402.

Vleeshouwers, L.M.; Veraghen, A. *Global Change Biol.*, 2002, 8, 519-530.

Visvanathan, C. In *Mechanical Biological Pre-Treatment of Solid Waste Prior to Landfill* International Conference on Integrated Solid Waste Management in Southeast Asian Cities: July, 5-7 July, Siem Reap, KH. 2005.

von Felde, D.; Doedens, H. In Sixth*International Waste Management and Landfill Symposium;*. In Christensen, T.H. Cossu, R., R. Stengmann, R. Eds. S. Margherita di Pula, IT, 1997.

Wang, W.J.; Baldock, J.A.; Dalal, R.C.; Moody, P.W. *Soil Biol. Biochem.*, 2004, 36, 2045-2058.

Weber, J.; Karczewska, A.; Drozd, M.; Licnar, S.; Jamroz, E.; Kokowicz, A. *Soil Biol. Biochem.* 2007, 39, 1294-1302.

Westlake, K. *Waste Manage Res.* 1997, 15, 453-461.

Wiemer, K.; Kern, M.; *Mechanical–biological Treatment of residual waste based on the dry stabilate method.* M.I.C. Baeza-Verlag: Witzenhonsen, (DE), 1995; pp.1-103.

Ximenes, F.A.; Gardner, W.D.; Cowie, A.L. *Waste Manage.* 2008, 28, 2344-2354.

Zach, A.; Binner, E.; Latif, M. *Waste Manage Res.* 2000, 18, 25-32.

Zimmerman, A.R.; Goyne, K.W.; Chorover, J, Komarneni, S.; Brantley, S.L. *Org Geochem.* 2004, 35, 355-375.

Zurbrügg, C. 2002. In Scientific Committee on Problems of the Environment (SCOPE) Urban Waste Management Review session. Durban, ZA, 2002.

Zwart, K. In: *Appying compost beneficts and needs*, 2001, Federal Ministry of Agriculture, Forestry, Environment and Water Management, Austria, and European Communities Eds, Brussels 23-23 November 2001, ISBN 3-902 338-26-1, 77-84, 2003.

In: Municipal Solid Waste
Editors: G. Konno and C. Machado
ISBN 978-1-62100-257-4

Chapter 6

USE OF INDUSTRIAL WASTE AND MUNICIPALITY SOLID WASTE AS AGGREGATE, FILLER OR FIBER IN CEMENT MORTAR AND CONCRETE

Nabajyoti Saikia and Jorge de Brito
Department of Civil Engineering and Architecture
Instituto Superior Técnico (IST), Technical University of Lisbon, Portugal

ABSTRACT

Management and treatment of industrial solid waste and municipal waste has been gaining importance. Waste material can be utilized as fuel in cement kilns, as raw material for cement clinker, as mineral addition to cement, or as granular material in cement mortar and concrete. Depending on the properties of the waste, it can be used in the cement and concrete industry without treatment or after treatment. Although vast amounts of waste material can be consumed in the production of cement clinker and blended cement, that consumption can be increased substantially if it is used as aggregate in cement mortar and concrete. Using waste material in this way can solve problems of shortage of aggregate in construction sites and reduce environmental problems related to aggregate mining and waste disposal.

For effective use of waste in the construction industry, a combined chemical, civil, material and environmental engineering approach is necessary. Careful characterization of waste is necessary before it is used in construction. Treatment of some waste materials is necessary to improve their quality before being incorporated into concrete. Environmental assessment of waste and concrete containing waste is also necessary for proper evaluation of waste materials as a constituent in concrete.

Slump, bulk density, unit weight, and air content; mechanical properties such as strength and elasticity modulus, and durability properties such as water absorption, shrinkage, carbonation and chloride resistance of fresh and hardened concrete containing waste materials, are normally evaluated. Some waste materials can improve certain concrete properties such as fire behaviour, thermal insulation and cracking, and therefore some reports are available on these topics too. But the durability performance of concrete containing waste materials has not been evaluated to the same extent and therefore significant research is necessary on this topic. As different waste materials can change the microstructure of mortar and concrete and the hydration chemistry of cement, more work is necessary to understand the aggregate-cement interaction in concrete and to develop more robust waste aggregate-based concrete technology.

In this chapter, the properties of plastic, municipal solid waste incineration ashes, coal bottom ash, tire rubber waste, different industrial slag and waste materials from miscellaneous industries when used as aggregate or filler or fibre in cement mortar and concrete are discussed. For proper presentation of the results the chapter is divided into subsections for each type of waste material. Finally, a concluding comment along with the scope of future research is presented.

Keywords: Waste materials, Concrete, Mortar, Aggregate, Fibre, Filler

1. INTRODUCTION

Management and treatment of industrial solid waste and municipal waste has been gaining importance. This waste ranges from relatively inert, e.g. glass bottles, excavated soil, building and demolition waste, to hazardous waste with high concentrations of heavy metals and toxic organic compounds. Waste material can be utilized as fuel in cement kilns, as raw material for cement clinker, as mineral addition to cement, or as granular material in cement mortar and concrete.

The production of cement and concrete is an energy-consuming process. Significant amounts of CO_2 are also emitted during the production of cement and concrete (Table 1). The use of waste materials in these operations can lower the amount of energy consumed and cut the amount of gases (including CO_2) released into the atmosphere.

Table 1. CO_2 emissions from cement and concrete production [Wilson 1993 (online)]

	Amount of CO_2 emitted in production per		% of total CO_2
	Ton of cement (lbs)	Cubic yard of concrete (lbs)	
Source of CO_2 emission from energy use	1410	381	60
Source of CO_2 emission due to limestone calcining	947	250	40
Total CO_2 emission	2410	631	100

Cement kilns can burn waste materials, such as used motor oil, spent solvents, printing inks, paint residue, cleaning fluids, and scrap tires, relatively more safely than an MSW incinerator because the extremely high temperatures in cement kilns result in very complete combustion with very low pollutant emissions [Knuttgen and Muench, no date (online), Wilson 1993 (online)]. The use of certain agricultural waste (e.g. peanut hulls) in cement kilns can also reduce CO_2 emissions [Wilson 1993 (online)]. Waste materials can also be used to make cement clinker. The high temperature processing of waste in the production of cement clinker can destroy toxic organic compounds without the formation of dioxins, fix metals in the product, and use mineral content as a constituent of clinker. Using some types of waste in clinker production may lower CO_2 emissions if the source of the calcium is different from $CaCO_3$. Extensive literature is available on the use of waste such as red mud, MSWI ash, steel mill scale, leather scraps and shavings, construction and demolition waste, and various sludges in the production of clinker [Caponero and Tenorio 2000, Espinosa and Tenorio 2000, Galbenis and Tsimas 2006, Monshi and Asgarani 1999, Saikia et al. 2007, Trezza and Scian 2007, Vangelatos et al. 2009].

Several waste materials are extensively used as mineral addition in the preparation of blended cement. Nowadays cements are produced using waste like fly ash (FA), ground granulated blast furnace slag (GBFS) [Taylor 1997]. The use of waste material as mineral addition in cement lowers the atmospheric emission of CO_2, reduces energy consumption and conveniently disposes of the waste.

Although vast amounts of waste material can be consumed in the production of cement clinker and blended cement, that consumption can be increased substantially if waste is used as aggregate in cement mortar and concrete. Using waste material in this way can solve problems of shortage of aggregate in construction sites and reduce environmental problems related to aggregate mining and waste disposal. The use of waste aggregate can reduce the cost of the concrete production, too.

Figure 1 shows a typical concrete composition. Aggregate typically accounts for 70% to 80% of the concrete volume and it plays a substantial role in concrete properties such as workability, strength, dimensional stability, and durability. Conventional concrete contains sand as fine aggregate and gravel, limestone and granite in various sizes and shapes as coarse aggregate. Interest is growing in using waste materials as aggregate, and significant research is underway on the use of granulated coal ash, blast furnace slag, fibre glass waste materials, waste plastics, rubber waste, sintered sludge pellets, and other materials as replacement for traditional aggregate.

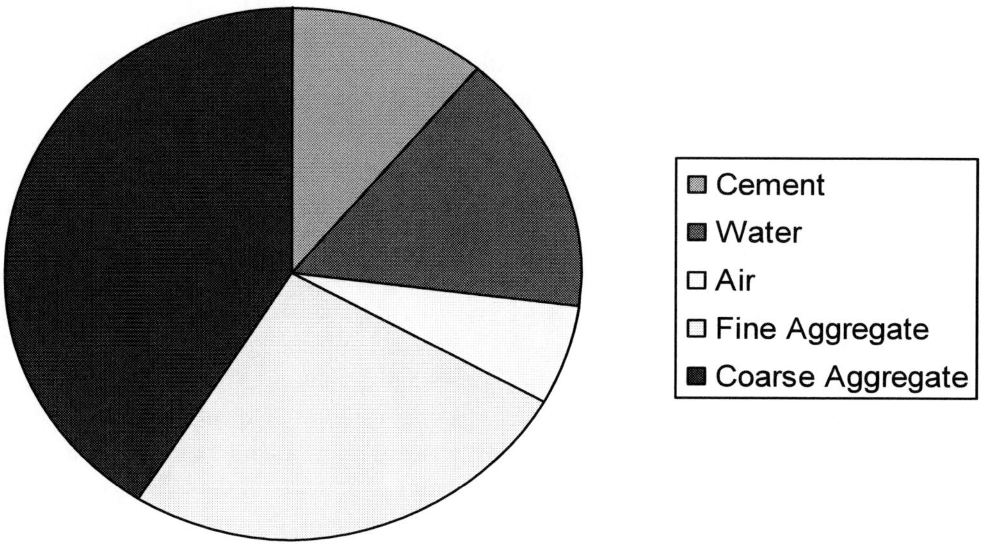

Figure 1. A qualitatively typical concrete mix.

Depending on its properties, waste can be used in the cement and concrete industry without treatment or after treatment. Some types of waste have to be treated because they contain undesirable components or perform poorly. Preliminary physico-chemical and mineralogical characterization of waste is therefore necessary before using the material in cement based systems. Waste materials may be mixed with other materials to prepare aggregate, too. Most industrial and municipal waste contains quite large amounts of toxic constituents, and therefore waste and concrete that contains waste should also be assessed for environmental impact as part of a proper evaluation of any waste materials that may be used as constituents in concrete. This indicates that chemical, civil, material and environmental engineering approches are necessary before waste can be used safely and effectively in the construction industry.

Although there is a great deal of literature on the utilization of different waste materials in the cement and concrete industry, this literature review will focus mainly on the utilization of plastic, municipal solid waste incineration ashes, coal bottom ash, tire rubber waste, various industrial slag and waste materials from other miscellaneous industries as aggregate, filler or fibre in cement mortar and concrete preparation.

2. Use of Plastic Waste in Concrete

Huge amounts of plastic solid waste are generated all over the world by the consumption of packaged products. Dumping plastic waste in landfills is not possible due to its slow degradation rate and bulky nature. The high calorific value of plastic waste can be used for incineration or other high temperature processes, but their combustion produces dangerous gases that could be harmful to human health. Although these alternatives are feasible, recycling waste plastic to produce new materials, such as concrete, appears to be the best solution for disposing of it since it is both cheaper and has ecological advantages.

There are quite a number of references on the use of plastic waste as an aggregate (mainly to replace fine aggregate) or as a fibre in the preparation of cement mortar and concrete [Siddique et al. 2009]. Figs. 2 and 3 are photographs of shredded plastic waste including a modified plastic waste (Figure 3(B)) shows those used to substitute part of the fine aggregate in the preparation of concrete.

Different types of plastic waste like polyethylene terephthalate (PET) bottle [Akcaozoglu et al. 2009, Albano et al. 2009, Choi et al. 2005, 2009, Kim et al. 2009, Marzouk et al. 2007, Silva et al. 2005, Yesilata et al. 2009], polyvinyl chloride (PVC) pipe [Kou et al. 2009], high density polyethylene (HDPE) [Naik et al. 1996], thermosetting plastics [Panyakapo and Panyakapo 2008], shredded plastic waste [Al-Manaseer and Dalal 1997, Ismail and Al-Hashmi 2008], expanded polystyrene (EPS) foam [Kan and Demirboga 2009a, 2009b], glass reinforced plastic (GRP) [Asokan 2009] are used as aggregate, filler or fiber in the preparation of cement mortar and concrete. Most of the studies are carried out at constant workability using standard methods. The slump, bulk density, unit weight, and air content of fresh concrete and different strength properties and elasticity modulus of hardened concrete are normally evaluated along with some special properties like fire behaviour, thermal insulation properties, microstructure and reactivity of plastic in alkaline solution.

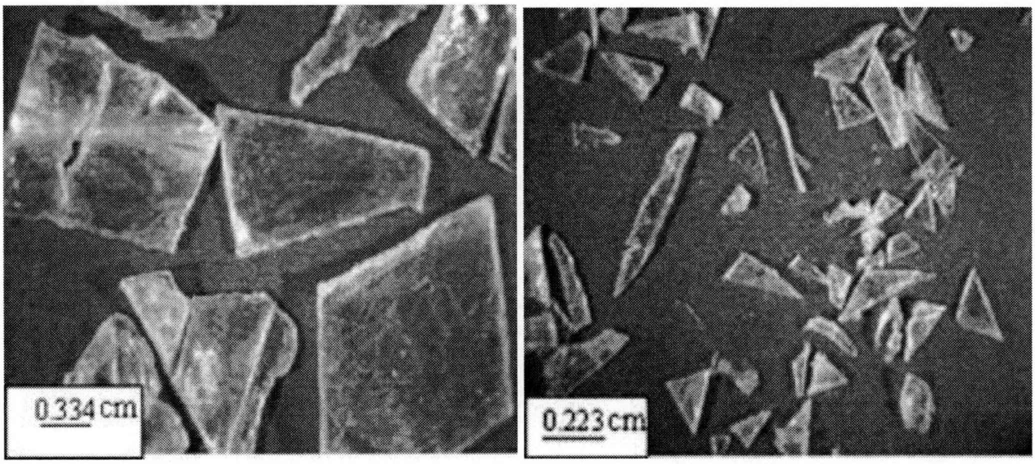

(A) (B)

Figure 2. Microphotograph of PET particles: (A) large; (B) small [Albano et al. 2009].

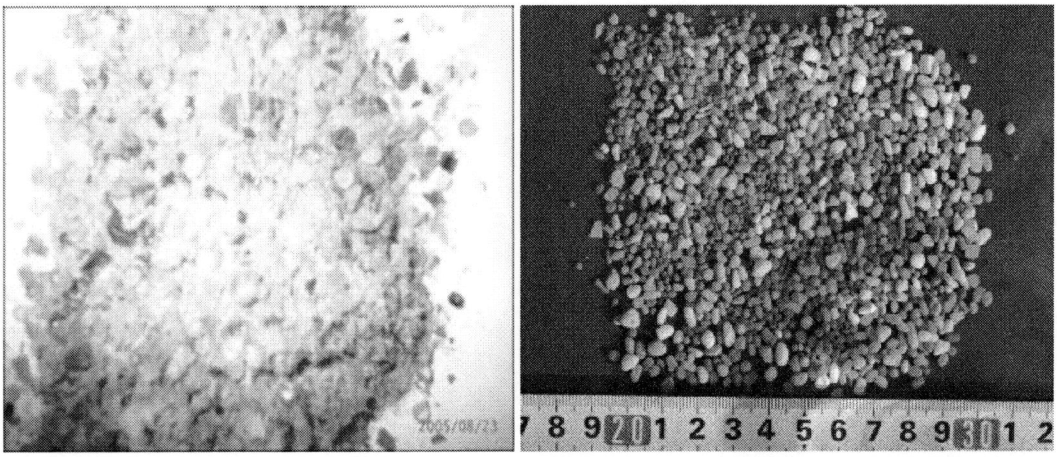

(A) (B)

Figure 3. Photographs of (A) waste plastic [Ismail and Al-Hashmi 2008] and (B) GBFS coated waste PET [Choi et al. 2005] used to replace some of the fine aggregate in concrete.

The incorporation of plastic in the form of aggregate or fiber lowers the slump value of fresh concrete because of the sharp edges and angular particle size of plastic aggregate (Figure 4) [Batayneh et al. 2007, Ismail and Al-Hashmi 2008, Soroushian et al. 2003]. However, an increase in slump value due to incorporation of this material has also been reported [Al-Manaseer and Dalal 1997, Bayasi and Zeng 1993, Choi et al. 2005, 2009]. Concrete mixes containing plastic aggregates will have more free water as plastic aggregates neither absorb nor add any water to the concrete mix and therefore this increases slump [Al-Manaseer and Dalal, 1997, Siddique et al. 2009]. Figure 5 shows the fresh density of concrete containing plastic as fine aggregate. The incorporation of plastic as aggregate decreases fresh and dry densities of resulting concrete [Ismail and Al-Hashmi 2008, Siddique et al. 2009].

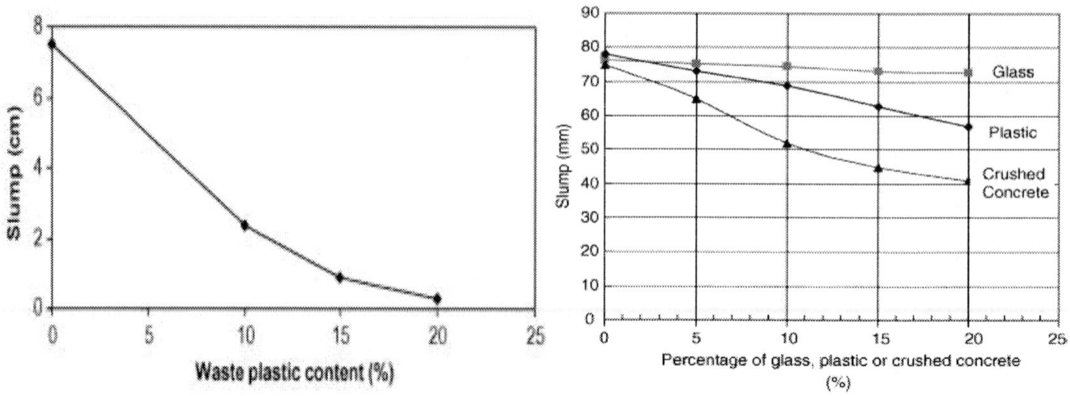

(A) (B)

Figure 4. Slump of concrete containing plastic waste as partial substitution of fine aggregate: (A) changes in slump due to increasing incorporation of plastic [Ismail and Al-Hashmi 2008] and (B) comparison of slump values of concrete containing glass and plastic waste as partial substitution of sand and crushed concrete as partial substitution of coarse aggregate [Batayneh et al. 2007].

The incorporation of plastic as aggregate decreases the modulus of elasticity and the compressive, splitting-tensile, and flexural strengths of the resulting concrete [Siddique et al. 2009]. Figs. 6 and 7 show the compressive and flexural strength of concrete containing plastic waste as partial substitution of fine aggregate. The factors that may be responsible for low compressive and flexural strength of concrete containing plastic are: 1. the decrease in bond strength between the surface of the plastic waste and the cement paste; 2. the hydrophobic nature of plastic waste, which can inhibit cement hydration reaction by restricting water movement. In a recent study it was found that the incorporation of plastic up to a certain level did not influence the compressive and flexural strength of concrete [Marzouk et al. 2007]. However, curing conditions were slightly different from those of the majority of the other studies and that might have had some effect on the properties.

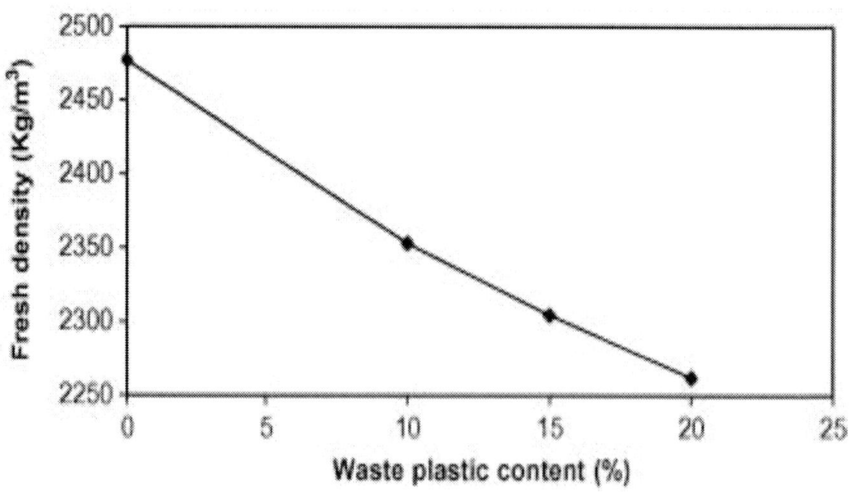

Figure 5. Fresh density of concrete due to increasing incorporation of plastic as fine aggregate [Ismail and Al-Hashmi 2008].

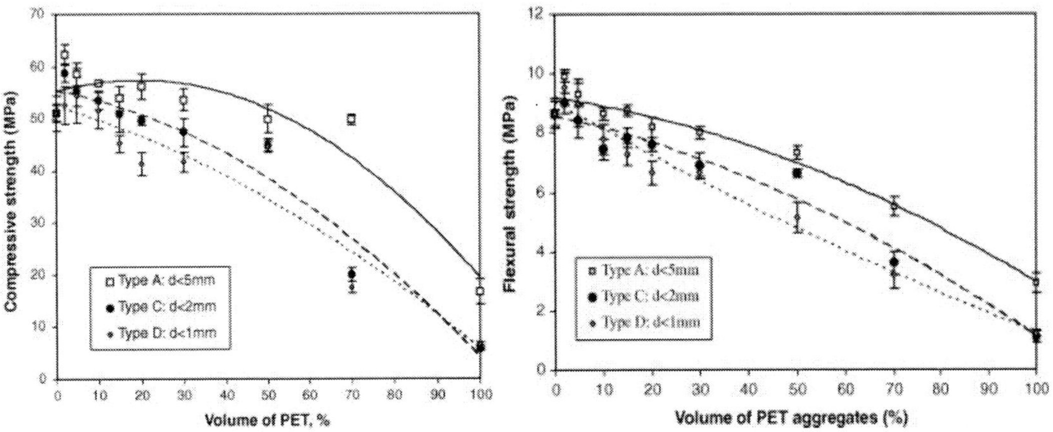

(A) (B)

Figure 6. Effect of recycled PET-aggregate on the compressive and flexural strength of the composites [Marzouk et al. 2007].

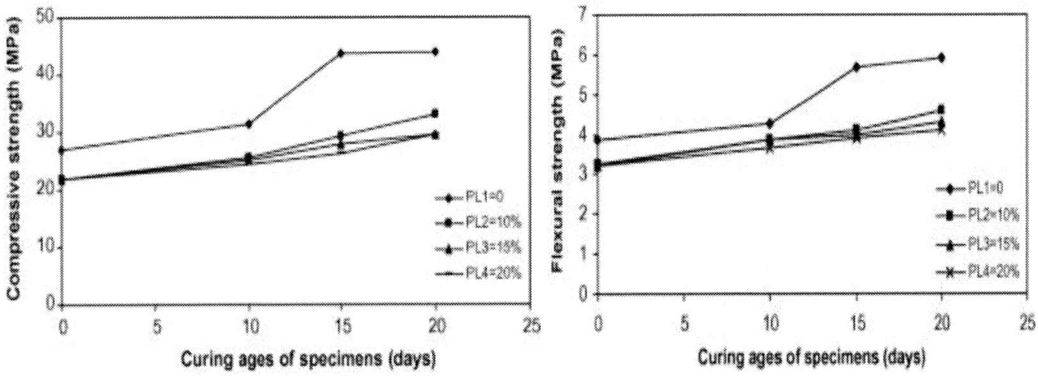

(A) (B)

Figure 7. Development of compressive and flexural strength of concrete containing plastic waste as fine aggregate for different substitution levels with respect to curing time [Ismail and Al-Hashmi 2008].

Concrete containing waste plastic showed no propagation of micro cracks and improved toughness compared with conventional concrete [Ismail and Al-Hashmi 2008]. However, toughness indices of concrete containing PET fiber decrease with longer curing time due to degradation of the fiber by alkaline hydrolysis [Silva et al. 2005]. The incorporation of plastic aggregate in plain concrete improves thermal insulation performance, which is dependent on the shape of the plastic aggregate, too [Yesilata et al. 2009].

The permeability of concrete is also improved when natural aggregate is replaced with plastic waste and with the incorporation of plastic as fiber [Akcaozoglu et al. 2009, Bayasi and Zeng 1993, Soroushian et al. 2003]. The inclusion of plastic up to a certain level does not influence the shrinkage of concrete [Akcaozoglu et al. 2009, Marzouk et al. 2007]. The sorptivity coefficient and rate of water adsorption of concrete containing PET-bottle as fine aggregate are lower than those of the reference concrete (Figure 8) [Marzouk et al. 2007]. The incorporation of EPS aggregate in concrete mixes improves their frost resistance. Coarse EPS aggregate is more susceptible to freeze–thaw cycles than fine EPS aggregate [Kan and

Demirboga 2009]. The incorporation of plastic fibers lowers the abrasion resistance and improves impact resistance of concrete [Bayasi and Zeng 1993, Soroushian et al. 2003]. However, only a few references are available on the evaluation of durability properties of concrete containing plastic waste aggregate.

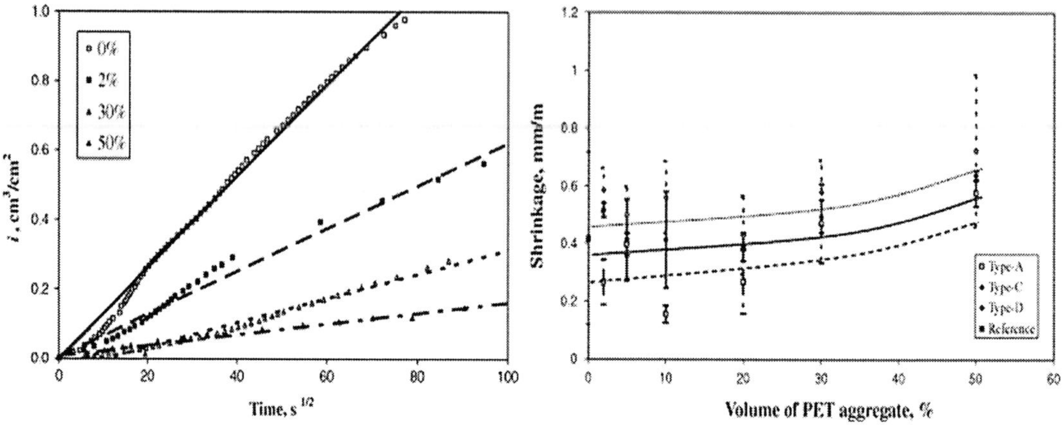

(A) (B)

Figure 8. (A) Sorptivity coefficient and (B) shrinkage (under conditions of complete saturation) of composites containing plastic aggregate [Marzouk et al. 2007].

(A) (B)

Figure 9. Transition zone between (A) cement paste and natural aggregate and (B) cement paste and GBFS coated PET aggregate in cement mortar after 28 days of curing [Choi et al. 2005].

The incorporation of mineral additions in cement may improve the properties of concrete containing plastic but not much is known except for a few reports. The incorporation of fly ash (FA) and blast furnace slag (GBFS) increases the compressive strength as well as the dry density of concrete containing plastic [Panyakapo and Panyakapo, 2008, Akcaozoglu et al. 2009]. Coating plastic aggregates with granulated blast furnace slag (GBFS) and sand powder can improve the bond between the plastic aggregate and the cement paste by forming hydration products [Choi et al. 2005, 2009]. Figure 9 shows the interfacial transition zone between cement paste and natural aggregate and between cement paste and GBFS coated PET aggregate in cement mortar after 28 days of curing. However, no comparative study has been done to evaluate the properties of concrete containing coated and uncoated plastic aggregates. Treating HDPE plastic with a solution containing 5% HOCl and 4% NaOH can improve the compressive strength of concrete containing HDPE as soft filler by improving bonding between the HDPE particles and the cement paste [Naik et al. 1996]. Heat treatment of EPS aggregate can considerably improve its density, strength and thermal conductivity [Kan and Demirboga 2009b].

Although a substantial number of references relate to the use of different types of plastic in concrete, further work is necessary to convert this type of waste into an important material for the preparation of concrete and to develop some specific types of concrete using its particular aggregate properties.

3. Municipal Solid Waste Incineration (MSWI) Ash

The MSW incineration process generates various types of waste, the major ones being bottom ash and fly ash. These ashes normally contain a large amount of toxic constituents and are therefore hazardous in nature (Table 2). The characteristics of ash are influenced by incineration parameters like furnace type and temperature, capacity, and waste input. MSWI ash is generally recycled in concrete after treatment or after solidification/stabilization in monolithic form, because it contains harmful components [Aubert et al. 2004, Cinquepalmi et al. 2008, Collivignarelli and Sorlini 2002, Pera et al. 1997]. The presence of harmful constituents like elemental Al, water soluble salts of chlorides and sulfates and organic compounds in MSWI ash, is the main problem with using this ash in concrete [Saikia et al. 2008].

Among the techniques used to improve MSWI ash properties in concrete are: combustion to remove organic compounds [Aubert et al. 2004]; water leaching or leaching at basic pH using chemical solutions to remove aluminium, soluble matter content, toxic elements, etc. [Aubert et al. 2004, Collivignarelli and Sorlini, 2002, Pera et al. 1997], and solidification by cement or other chemicals to immobilize toxic components of MSWI ashes [Aubert et al. 2004, Cinquepalmi et al. 2008]. Weathering may be a method of treating MSWI ashes because atmospheric carbonation of MSWI ash can immobilize some toxic elements, too [Chimenos et al. 2000]. The evaluation of physical-chemical, mineralogical and environmental characteristics of waste and mechanical and durability-related properties, along with the environmnetal suitability of concrete, is therefore necessary for the proper utilization of MSWI ashes in concrete mixes.

All techniques can improve the mechanical properties of concrete containing MSWI ashes but these properties of ash-containing concrete are always inferior to the properties of conventional concrete and therefore not suitable for structural concrete preparation [Aubert et al. 2004, Cinquepalmi et al. 2008, Collivignarelli and Sorlini, 2002, Gines et al. 2009, Pera et al. 1997]. Figs. 10 and 11 show the compressive strength of cement mortar containing cement solidified MSWI fly ash and weathered MSWI bottom ash aggregates respectively. The leaching of toxic elements from concrete containing MSWI ash is within the standard specification and is, in some cases, almost the same as conventional concrete (Figure 12).

Several references can be found on the use of MSWI ashes in concrete but a clear picture is not available on the use of these ashes as aggregate in concrete.

Table 2. Total and leached amounts of some elements from two MSWI ash samples [Saikia et al. 2008]

Elements	MSWI ash 1		MSWI ash 2		Limit*
	Total Amount	Leached Amount	Total Amount	Leached Amount	
As	na	na	312	**16.4**	0.80
Cd	14.06	**0.07**	30.15	**0.17**	0.03
Cr	292	**18.5**	67.6	**1.8**	0.50
Cu	3555	**16.10**	3140	0.14	0.50
Mo	29.5	**1.8**	31.0	**2.6**	1.5
Ni	217	0.37	129	0.35	0.75
Pb	3518	0.32	1126	0.26	1.3
Sb	3077	**3.4**	1208	0.60	2.0
Se	292	na	26.4	**0.31**	0.10
V	805	0.15	252	0.12	32
Zn	5905	0.39	2997	0.83	2.8

na: not analysed.

* Limit values for recycling of waste as construction material in Flanders region, Belgium according to the Flemish column test NEN 7343.

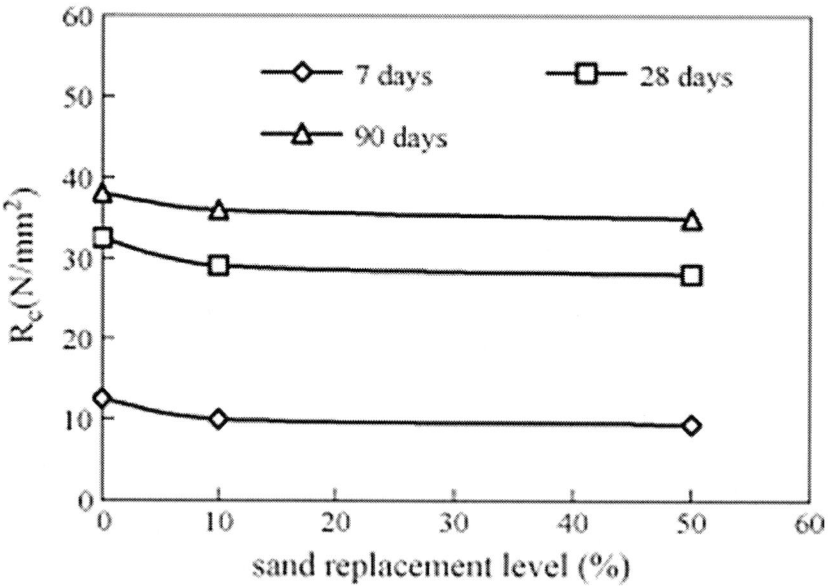

Figure 10. Development of compressive strength of cement mortar with replacement of aggregate by cement sodified MSWI fly ash [Cinquepalmi et al. 2008].

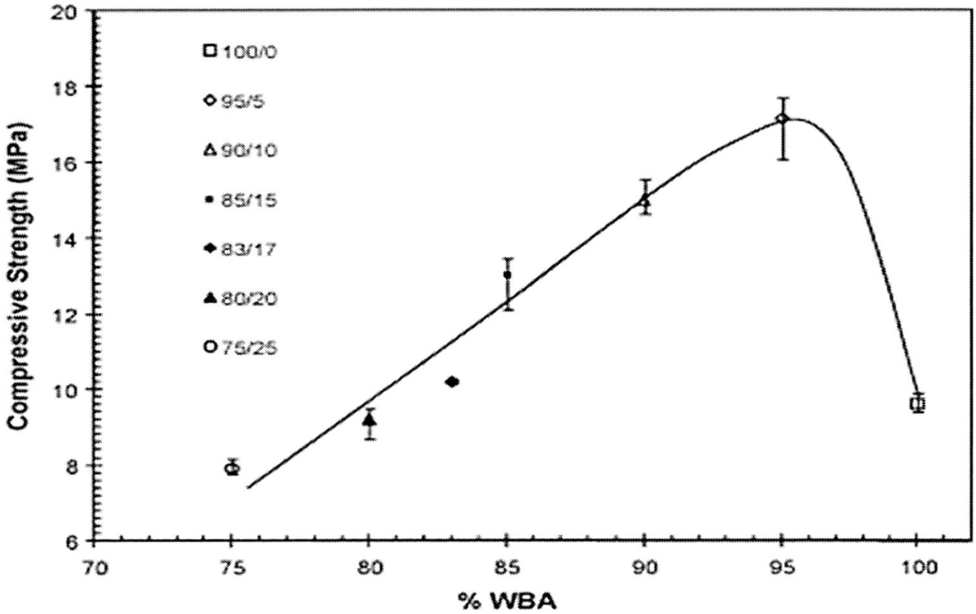

Figure 11. Variation in compressive strength of cement mortar with percentage of weathered MSWI bottom ash (WBA) content (100/0 means 100% WBA aggregate containing mortar, 75/25 means a aggregate mixture of 75% WBA and 25% MSWI fly ash containing mortar) [Gines et al. 2009].

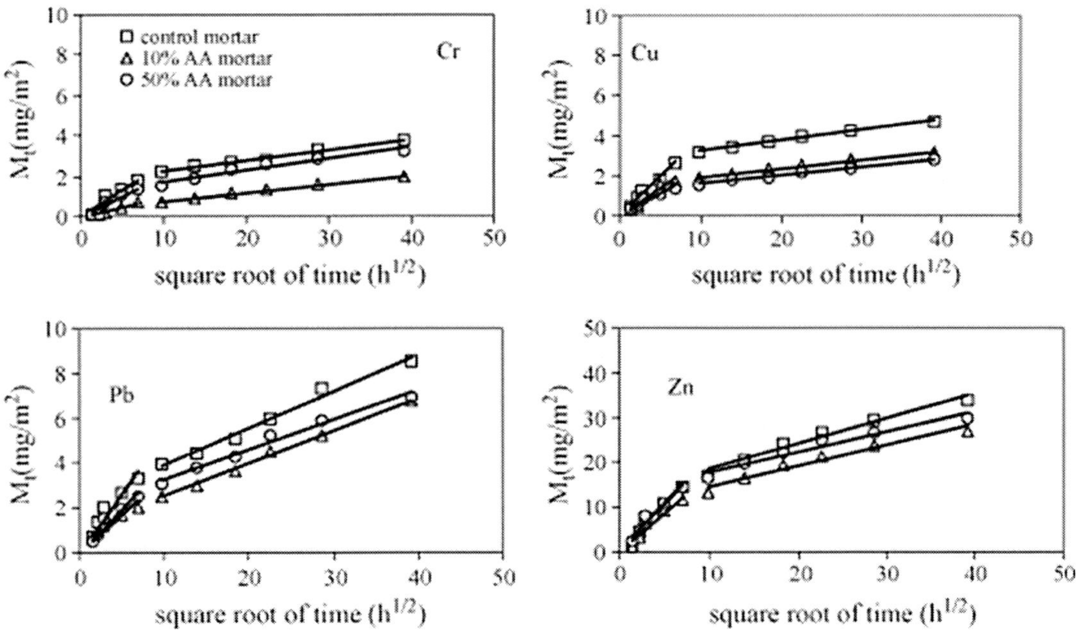

Figure 12. Leaching of some toxic elements from cement mortar containing cement sodified MSWI fly ash as aggregate [Cinquepalmi et al. 2008].

As MSWI waste is hazardous in nature (like most industrial waste) a different methodology is needed to properly evaluate these materials. Present research results suggest that MSWI ashes should only be used in non-structural concrete. Evaluation of durability properties and long term mechanical behaviour of MSWI ash-containing concrete is needed to properly use this material. Further work is necessary to understand the effect of the chlorides and sulfates in MSWI ashes on the behaviour of concrete containing these ashes. A set of technical requirements for MSWI ashes used in concrete and for concrete containing MSWI ashes is needed, along with recommendations and regulations on the industrial production of concrete and other building products containing MSWI ash.

4. Coal Ash as an Aggregate in Concrete

Though a significant number of studies are available on the properties and use of fly ash as a mineral addition in normal Portland cement, not much literature exists on the use of fly ash and coal bottom ash (CBA) as a granular additive in concrete. The CBA falls into the furnace bottom in modern large thermal power plants and constitutes about 20% of the total ash content of the coal fed into the boilers. The properties of bottom ash depend on the coal type, pulverizing system, combustion conditions, temperature, type of furnace, minerals in the coal, and milling system. It is collected from the bottom of the combustion chamber in a solid granular form.

Figure 13 shows the X-ray diffraction (XRD) pattern of typical low calcium CBA. The XRD pattern of CBA indicates the presence of a glassy phase with strong peaks of mullite and quartz. Unlike its companion, pulverised fuel ash (PFA), it usually has low pozzolanic property which makes it unsuitable for use as a mineral addition in cement. It may contain higher concentrations of unburned carbon. Some power plants recover coal mill rejects with the bottom ash and therefore CBA may contain the pyrites that come from mill rejects, which can cause expansion within the concrete. The salt content and low pH in bottom ash may be corrosive and therefore evaluation of the corrosive nature of the bottom ash being used should be investigated. The corrosivity indicator tests normally used to evaluate bottom ash are pH, electrical resistivity, soluble chloride content, and soluble sulfate content. However, the fused and glassy texture of CBA (Figure 13) along with particle size (Figure 14) would normally make it an ideal substitute for the sand fraction of concrete [Babcock and Wilcox 1978].

Although similar to natural fine aggregate, bottom ash is lighter and more brittle and more closely resembles cement clinker [Rogbeck and Knutz 1996]. Environmental assessment of bottom ash did not identify environmental hazards associated with the beneficial use of this material [Tandon and Picornell 1998]. An important use of bottom ash is as a base layer material in road construction [ACAA 2007].

The chemical property of CBA is normally controlled by the properties of coal (origin of coal). Table 3 shows the oxide composition of some CBA samples, along with some physical properties. CBA is composed primarily of silica (SiO_2), ferric oxide (Fe_2O_3), and alumina (Al_2O_3), with smaller quantities of calcium oxide (CaO), potassium oxide (K_2O), sodium oxide (Na_2O), magnesium oxide (MgO), titanium oxide (TiO_2), phosphorous pentoxide (P_2O_5), and sulfur trioxide (SO_3).

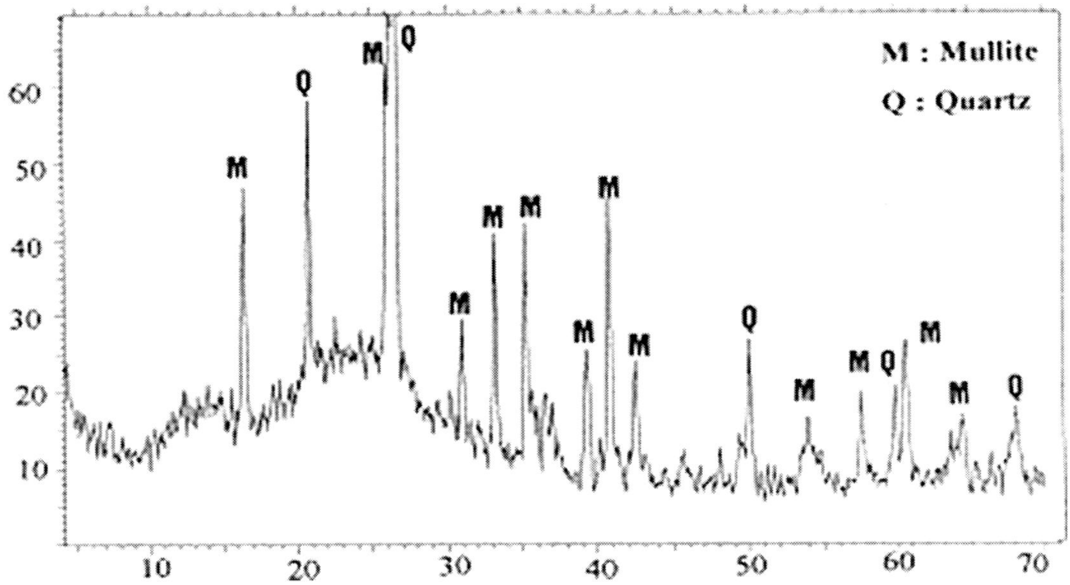

Figure 13. Mineralogy of a typical low calcium CBA [Cheriaf et al. 1999].

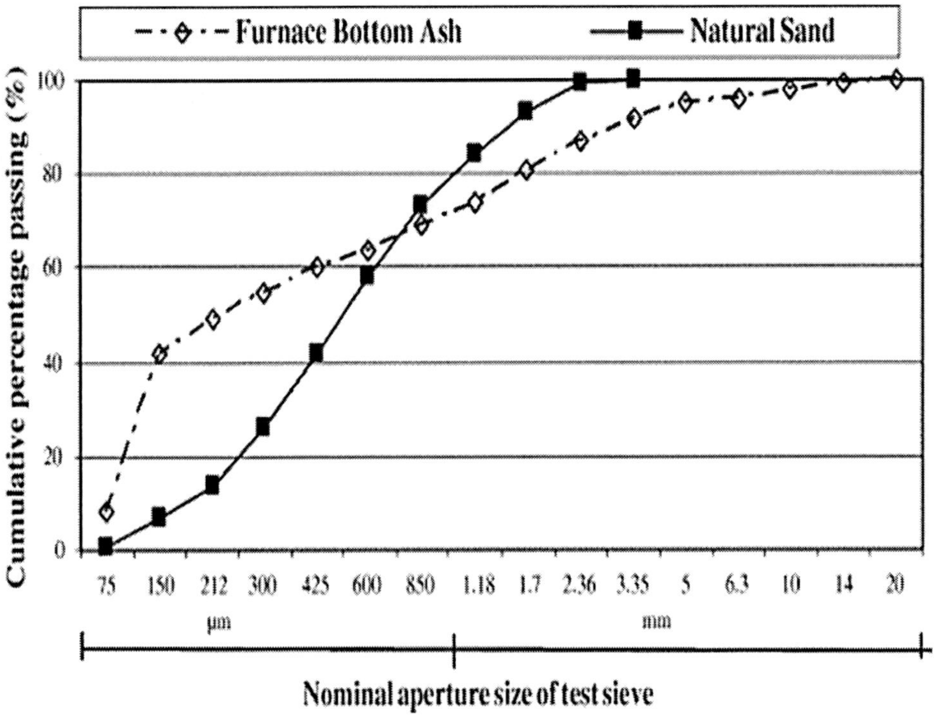

Figure 14. Particle size distribution of a CBA and natural sand [Bai et al. 2005].

Table 3. Chemical and physical characteristics of Portland cement and bottom ash

Content (%)	Andrade et al. 2007	Bai et al. 2005	Cheriaf et al. 1999	Kou and Poon 2009	Lee et al. 2010	Ozkan and Yuksel 2007	Kurama et al. 2009	Park et al. 2009	Ghafoori and Bucholc 2006
Country	Brazil	UK	France	China	S. Korea	Turkey	Turkey	S. Korea	USA
	nr	nr	Low-grade coal	nr	nr	nr	nr	nr	Lignite
Chemical analysis									
SiO_2	50.46	61.80	56.00	60.7	44.2, 48.0	59.53	54.5	47.9	41.70
Al_2O_3	28.35	17.80	26.70	18.3	31.5, 31.3	20.12	15.4	25.94	17.10
Fe_2O_3	10.69	6.97	5.80	6.56	8.87, 8.26	13.08	11.16	4.76	6.63
Na_2O	nr	0.95	0.20	0.89	nr	-	nr	1.38	1.38
K_2O	3.81	2.00	2.60	2.12	4.04, 4.14	0.06	1.34	0.67	0.40
CaO	2.07	3.19	0.80	3.25	2.00, 1.41	2.02	4.69	2.48	22.50
TiO_2	1.57	0.88	1.30	0.95	2.38, 2.29	nr	nr	0.86	nr
SO_3	0.34	0.79	0.1 (as S)	0.82	1.3, 1.2	trace	1.30	1.03	0.42
ZrO_2	0.18	nr	nr	nr	nr	nr	nr	nr	nr
V_2O_5	0.09	nr	nr	nr	0.07, 0.09	nr	nr	nr	nr
MnO	0.07	nr	nr	nr	0.07, 0.06	nr	nr	nr	nr
ZnO	0.03	nr	nr	nr	nr	nr	nr	nr	nr
Y_2O_3	0.03	nr	nr	nr	nr	nr	nr	nr	nr
SrO	0.03	nr	nr	nr	nr	nr	nr	nr	nr
MgO	nr	1.34	0.60	1.28	2.6, 2.0	3.20	4.26	1.10	4.91
Cr_2O_3	nr	nr	nr	nr	0.03, 0.05	nr	nr	nr	nr
P_2O_5	nr	0.20	nr	nr	0.3, 0.33	nr	nr	nr	nr
Others	nr	0.49	nr	1.00	0.32, 0.40	nr	nr	nr	3.83
Cl	nr	nr	nr	nr	2.32, 0.47	nr	nr	0.18	nr
Loss on ignition	2.30	3.61	4.60	4.13	nr	9.81	8.90	3.63	1.13

Minerals									
nr	Quartz, mullite, glassy phase	nr	Mullite, quartz, glassy phase	nr	nr	nr	Quartz, magnetite, hematite, glassy phase	nr	nr
Physical properties									
Specific gravity (g/cm^3)	1.674	1.50-1.58	nr	nr	1.97,2.01	1.39	nr	nr	2.33 (OD Cond.) 2.47 (SSD Cond.)
Water absorption (%)	nr	30.4-32.2	nr	nr	3.61, 0.56	6.10	nr	2.11-2.43	7.0
Clay lump (%)	nr	nr	nr	nr	nr	nr	nr	nr	9.8

nr: not reported.

Figure 15. Photographs of different size fractions of a typical CBA: (a) 0-2 mm; (b) 2-8 mm and (c) 8-11 mm [Lee et al. 2010].

Figure 15 shows the different particle size fractions for typical CBA use as a partial substitution of fine and coarse aggregate in concrete. Figure 16 shows the scanning electron micrographs of porous CBA and natural sand. The use of bottom ash in concrete presents several technical challenges: its physical and mineralogical characteristics; the effect on water requirement and the part played in cement hydration. Two different approaches were adopted to prepare mixes of concrete containing CBA: one considering the natural humidity present in the porous particles and the other not considering it, but seeking to maintain the same strength. Also, bottom ash has a porosity with a high capacity to retain water, a factor which should be considered in the evaluation of the concrete produced with this byproduct since bottom ash works well as a water reservoir.

At fixed W/C value, the slump value of concrete with CBA is higher than that of concrete without CBA and it increases with the replacement ratio of sand by CBA (Figure 17) [Bai et al. 2005, Kou and Poon 2009]. The water loss due to bleeding of concrete with CBA is higher than that of conventional concrete [Andrade et al. 2007, 2009, Ghafoori and Bucholc 1996] and both parameters increase with higher CBA content in the concrete [Andrade et al. 2007, 2009]. The setting time of concrete is also effected by the presence of CBA in concrete and setting takes longer for concrete containing CBA prepared at higher W/C value [Andrade et al. 2009] and prepared at constant slump [Ghafoori and Bucholc 1996]. At fixed slump, the plastic shrinkage and unit weight of the concrete containing CBA as fine aggregate are lower than those of the conventional concrete which contains natural sand [Ghafoori and Bucholc 1996]. At fixed slump, the air content of concrete containing CBA as fine aggregate is almost similar to that of the concrete containing natural sand as aggregate [Ghafoori and Bucholc 1996].

Figure 16. Scanning electron micrographs of CBA (a, b, c, d) and normal sand (e, f) [Lee et al. 2010].

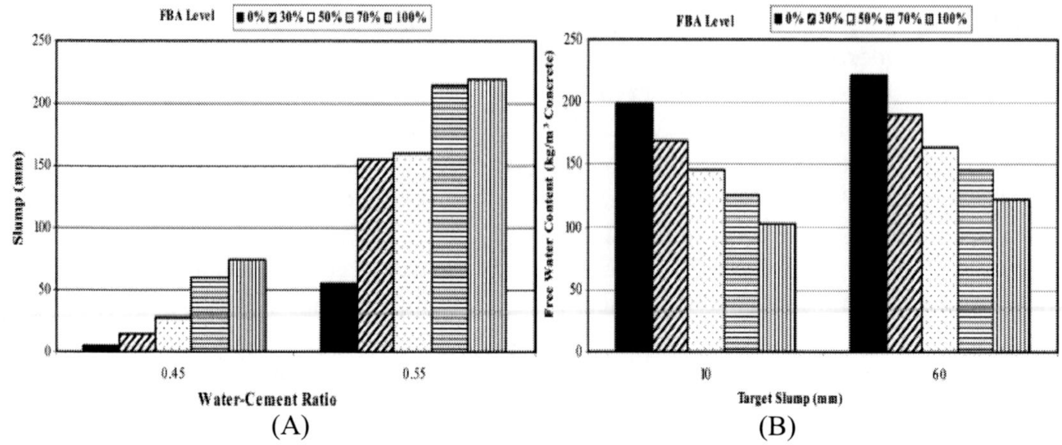

Figure 17. Slump and free water content of concrete containing CBA as fine aggregate: (A) slump value at fixed W/C; (B) free water content at fixed slump range [Bai et al. 2005].

The compressive strength of concrete containing CBA and combined CBA and natural sand mixes is lower than that of the concrete containing normal sand prepared at constant W/C value [Agarwal et al. 2007, Bai et al. 2005] and almost comparable at constant workability [Bai et al. 2005, Kou and Poon 2009] (Figure 18). The compressive strength of the concrete containing CBA and CBA and natural sand mixes, prepared by discounting the natural water content in bottom ash, is similar to or even better than that of concrete containing normal sand [Andrade et al. 2007].

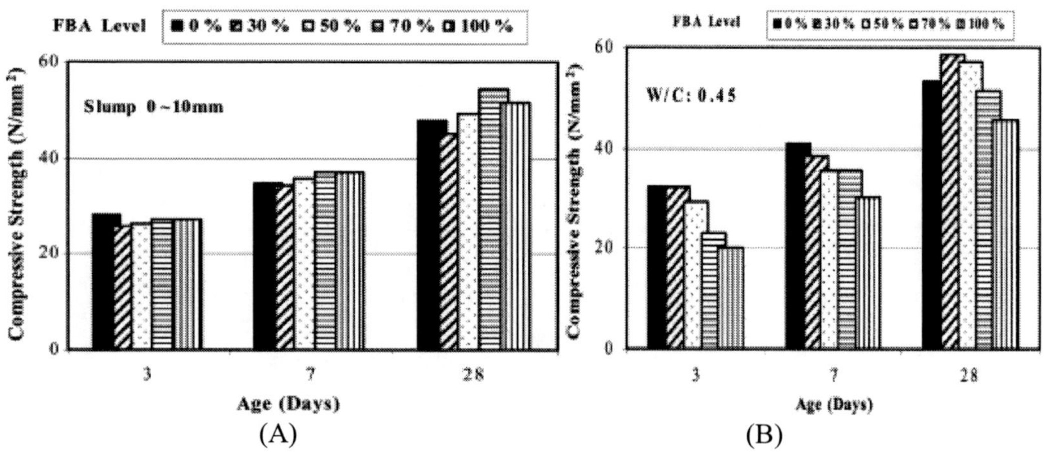

Figure 18. Compressive strength of concrete containing CBA: (A) at fixed slump; (B) at fixed W/C [Bai et al. 2005].

Table 4 shows the sorptivity coefficient of concrete containing CBA as fine aggregate at different substitution levels and of the control concrete specimen (0%), determined from capillary absorption of water and ethanol. Results indicate that the sorptivity coefficient and therefore the capilliary water absorption of concrete containing CBA is higher than that of conventional concrete [Andrade et al. 2007, 2009]. The air permeability, sorptivity and drying shrinkage of concrete increase with increasing replacement of sand by CBA [Basheer and Bai

2005]. The depth of carbonation in concrete increases with increasing replacement of sand by CBA [Basheer and Bai 2005, Kasemsaisiri and Tantermsirikul 2008].

Table 4. Sorptivity coefficient of concrete containing CBA as fine aggregate at different substitution levels [Andrade et al. 2007]

Concrete	Sorptivity coefficient ($kg/m^2/h^{1/2}$)									
	Water					Ethanol				
	0%	25%	50%	75%	100%	0%	25%	50%	75%	100%
CRT3	2.2	3.1	3.8	5.7	8.9	1.1	1.4	1.8	2.8	4.0
CRT4		2.3	3.2	4.4	5.3		1.2	1.4	1.9	2.2

CRT3: Concrete mixture containing CBA, considering the natural water content of CBA, without discounting the water; CFT4: Concrete mixture containing CBA, discounting the natural water content, considering all (bottom ash + natural moisture) as solid material.

Drying shrinkage of concrete containing CBA and natural sand mixes is lower for concrete containing natural sand prepared at fixed W/C value and higher at fixed slump (Figure 19) [Bai et al. 2005]. However, at constant W/C, higher drying shrinkage of self-compacting concrete prepared with partial replacement of sand by CBA than that of concrete containing only sand is also reported [Kasemsaisiri and Tantermsirikul 2008]. Replacement of fine aggregate by CBA does not effect the mass loss due to abrasion but it does improve the resistance to chloride ingress and salt scaling [Basheer and Bai 2005]. At fixed W/C value, the resistance of concrete briquette to freeze–thaw decreases as the replacement of sand by CBA increases [Yuksel and Bilir 2007].

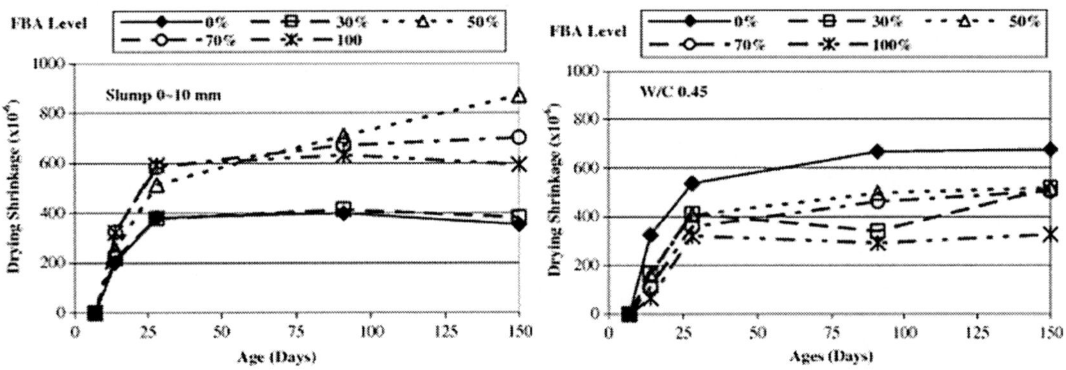

(A) (B)

Figure 19. Drying shrinkages of concrete containing CBA: (A) at fixed slump; (B) at fixed W/C [Bai et al. 2005].

At fixed slump, concrete with CBA as fine aggregate shows better resistance to chloride ion penetration than conventional concrete, due to the pozzolanic property of CBA [Kou and Poon 2009]. But at fixed W/C the resistance to chloride ion penetration deteriorates with an increasing amount of bottom ash in concrete [Kasemsaisiri and Tantermsirikul 2008, Kou and Poon 2009]. Replacement of sand with bottom ash improves the sulphate resistance of the resulting concrete and increasing the proportion further improves this property of concrete

[Kasemsaisiri and Tantermsirikul 2008]. The incorporation of GBFS with CBA to partially replace fine aggregate reduces the compressive strength of the resulting concrete. But the incorporation of FA with cement can improve the strength of concrete containing GBFS and CBA [Ozkan et al. 2007].

The workability of concrete mixes containing fly-ash as partial substitution of sand is better than that of concrete prepared without fly ash. The water requirement of concrete mixtures containing fly-ash as partial substitution of sand is slightly higher than for concrete prepared without fly ash. The rate and volume of water loss due to bleeding from concrete containing fly ash are similar to those of concrete prepared without fly ash. The setting of the concrete mix is delayed due to the incorporation of fly-ash in concrete as partial replacement of fine aggregate [Ravina 1997]. The incorporation of fly ash as sand replacement improves strength properties and abrasion resistance of concrete, due to the pozzolanic action of fly-ash, which improves interfacial bond between paste and aggregates [Siddique 2003a, 2003b].

5. Use of Rubber Tire as Aggregate or Fibre or Filler in Concrete

The disposal of rubber tire waste has become a serious problem for modern civilization because of the huge amounts of tires that are discarded, which are non-biodegradable in nature. The accumulation of used tires at landfill sites also presents the threat of uncontrolled fires which would produce a complex mixture of chemicals that could contaminate soil and vegetation and also provide breeding grounds for rats, mice, vermin and mosquitoes.

Innovative solutions to meet the challenge of the tire disposal problem are urgently needed because of the rapid depletion of available sites for waste disposal. Tire rubber can be used in asphaltic concrete mixtures, in incinerators to produce steam, to produce different plastic and rubber products, as a fuel for cement kiln, as feedstock for making carbon black, and as artificial reefs in marine environment, to name some of the attractive utilization options [Siddique and Naik 2004]. There is extensive literature, including excellent reviews, on the use of tire rubber as coarse or fine aggregate or as a filler material for the preparation of different types of concrete [Kumaran et al. 2008, Siddique and Naik 2004]. Some results of concrete containing rubber tire waste particles as aggregate or fiber or filler material are highlighted here:

- Increasing incorporation of rubber as aggregate, fiber or filler in concrete lowers the workability and bleeding of concrete [Pierce and Blackwell 2003, Hernandez-Olivares and Barluengab 2004]; concrete containing rubber as aggregate has acceptable workability in terms of ease of handling, placement, and finishing [Khaloo et al. 2008, Raghvan et al. 1998];
- The incorporation of rubber increases the water requirement of the resulting cement mortar composition [Yilmaz and Degirmenci 2009]; because of the low specific gravity of rubber aggregates the resulting concrete containing rubber has lower unit weight than conventional concrete [Siddique and Naik 2004];

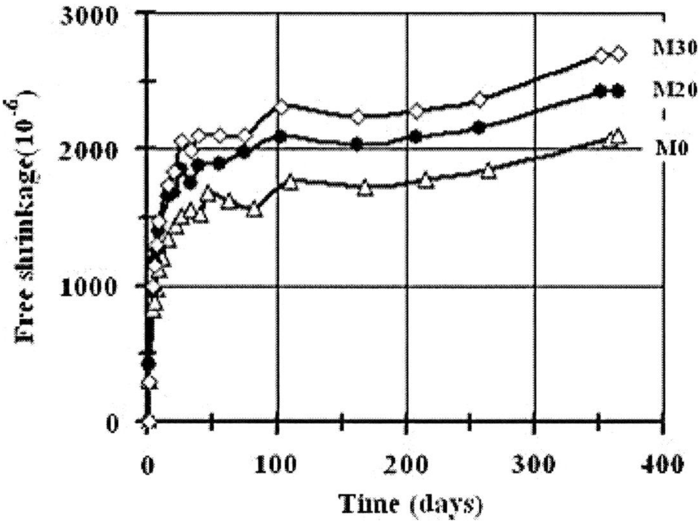

Figure 20. Influence of rubber aggregate substitution on free shrinkage of cement mortar (0%, 20% and 30% of the aggregate total volume is replaced by rubber aggregate in M0, M20 and M30 mortars, respectively [Turatsinze et al. 2007].

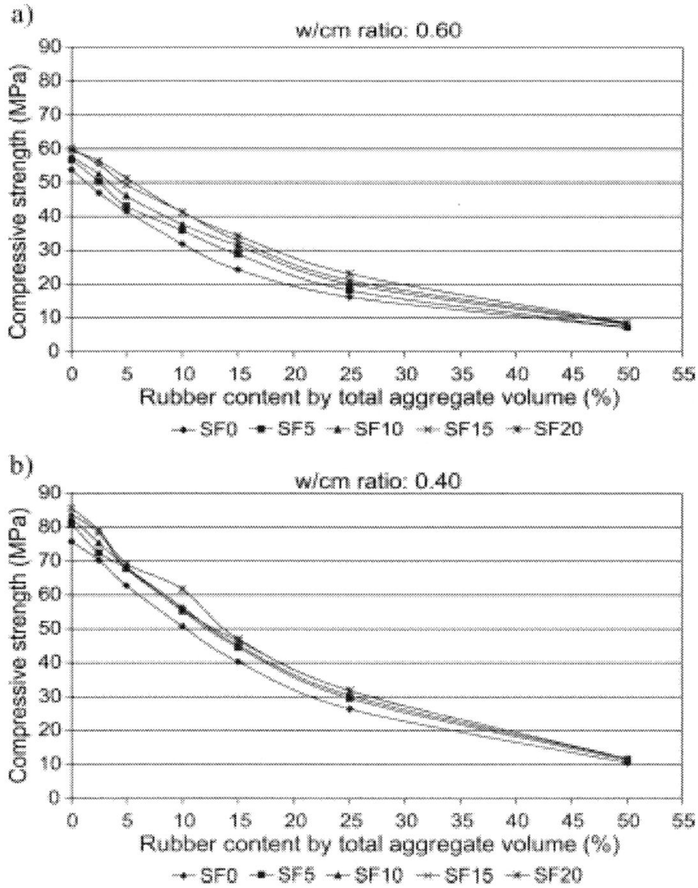

Figure 21. Variation in compressive strength of concrete containing rubber and silica fume for different water to cement ratios [Guneyisi et al. 2004].

- The air content of concrete that contains rubber is higher than that of conventional concrete, which is due to the non-polar nature of rubber, and its tendency to trap air in its rough surfaces and to attract air due to its water repellent nature [Siddique and Naik 2004];
- The incorporation of rubber granules and shreds into cement mortar reduces plastic shrinkage cracking and also delays the onset of cracking in comparison to control mortar [Raghvan et al. 1998]; incorporation of rubber in concrete increases free shrinkage length changes (Figure 20) [Turatsinze et al. 2007];
- The incorporation of rubber decreases the compressive and splitting tensile strength and elastic modulus and increases the flexural strength of concrete (Figure 21) [Guneyisi et al. 2004, Kumaran et al. 2008, Siddique and Naik 2004, Turatsinze et al. 2007, Yilmaz and Degirmenci 2009]; a decrease in the flexural strength of cement mortar and concrete containing rubber tire as aggregate is also reported [Kang and Jiang 2008]; the compressive strength of concrete containing coarse rubber aggregate is lower than that of the concrete containing fine rubber aggregate [Benazzouk et al. 2003, Eldin and Senouci. 1993, Khatib and Bayomy 1999, Topcu 1995]; however, the opposite trend for the compressive strength behaviour of concrete containing coarse and fine rubber particles is also reported [Fattuhi and Clark 1996, Ali et al. 1993]; the addition of silica fume to normal Portland cement improves the strength properties of concrete containing rubber tire aggregate (Figure 21) [Guneyisi et al. 2004]; for the same unit weight, the strength properties of concrete containing expanded rubber aggregate are better than those of concrete containing compact rubber aggregate [Benazzouk et al. 2003];
- Pretreatment of rubber waste with water, using chemical solutions like NaOH solution, carbon tetrachloride solution to improve the strength properties of concrete that contains rubber particles as granular additive or addition into cement, is reported by various authors [Eldin and Senouci. 1993, Li et al. 2004, Rostami et al. 2000, Segre and Joekes 2000]; treatment of rubber powder using NaOH and modification of rubber particles with latex can improve the bond between rubber particles and cement paste [Segre and Joekes 2000, Lee et al. 1998]; but treatment of rubber chips using NaOH does not improve the binding properties of rubber and cement paste and therefore it does not improve the properties of concrete that contains treated rubber chips [Li et al. 2004];
- The use of rubber as aggregate in concrete containing magnesium oxychloride cement gives higher strength and better rubber particle-cement paste binding characteristics than those exhibited by the Portland cement based concrete containing rubber particles [Siddique and Naik 2004];
- The addition of rubber as aggregate or shreds can improve the toughness (expressed in a toughness index), defined as a ratio between the area under the stress-strain curve up to 80% of the ultimate stress and the area under the same curve up to the ultimate stress of concrete [Benazzouk et al. 2007, Khaloo et al. 2008, Siddique and Naik 2004]; but if too much is incorporated this may lower the toughness of concrete due to low compressive strength [Khaloo et al. 2008];
- The incorporation of rubber as aggregate improves cracking behaviour and delays the cracking time of the resulting cement mortar [Benazzouk et al. 2007, Kang and Jiang

2008]; concrete containing rubber exhibits lower stiffness and a less brittle nature than conventional concrete (Figure 22) [Hernandez-Olivares and Barluengab 2004, Sukontasukkul and Chaikaew 2006];

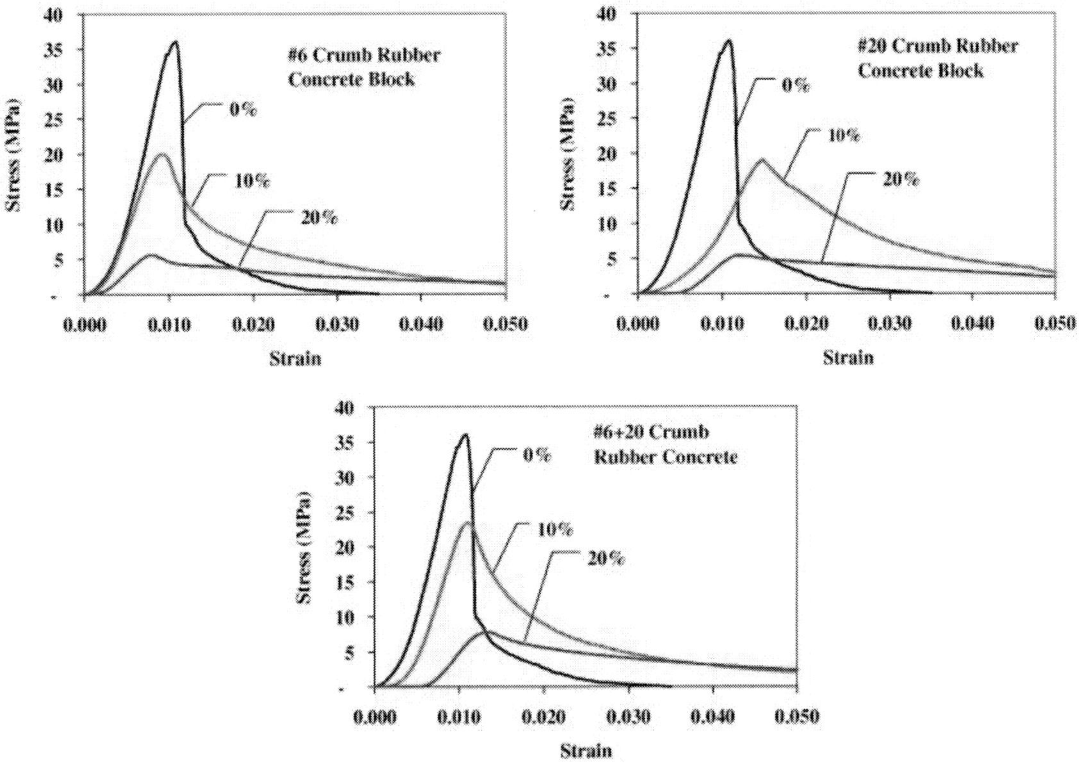

Figure 22. Compressive loading response of crumb rubber concrete blocks [Sukontasukkul and Chaikaew 2006].

- Concrete that contains rubber can absorb sound and vibration energy [Khaloo et al. 2008]; the heat insulation properties of concrete are also improved with the addition of rubber particles [Yesilata et al. 2009];
- The impact resistance of concrete with rubber as aggregate or filler is comparable to or better than that of a conventional concrete mix owing to the ability of the material to absorb energy [Eldin and Senouci 1993, Fattuhi and Clark 1996, Topcu 1995];
- The incorporation of rubber as fiber reduces the risk of explosive spalling of high strength concrete when it is exposed to high temperature [Hernandez-Olivares and Barluengab 2004];
- The capillary water absorption and water sorptivity of cement mortar that contains rubber as fine aggregate is lower than for normal mortar, and capillary water absorption and sorptivity decrease with increasing rubber particle incorporation [Benazzouk et al. 2007]; the incorporation of rubber as coarse aggregate or filler increases the water absorption and decreases the water permeability depth of the resulting concrete [Ganjian et al. 2009];

- The incorporation of rubber in concrete improves its freezing and thawing resistance [Siddique and Naik 2004];
- Concrete pavement blocks containing rubber crumbs as aggregate exhibit better skid resistance and less abrasion resistance than conventional concrete pavement blocks (Figure 23) [Sukontasukkul and Chaikaew 2006]; treatment of rubber particles using NaOH solution can improve the abrasion resistance of concrete containing rubber (Figure 24) [Segre and Joekes 2000];
- At constant compressive strength, the tensile strength, flexural strength and ultimate tension elongation of roller-compacted concrete increases with increasing substitution of fine aggregates by rubber particles [Kang et al. 2009].

Because of the improvement of some specific properties such as ductility, energy absorption efficiency, shrinkage and cracking associated with hardening and curing, concrete containing rubber as aggregate can be used for various purposes [Siddique and Naik 2004]. However, a major drawback of using rubber in concrete is the deleterious effect of rubber on the strength behaviour of concrete, and therefore further research is necessary to improve the strength and related behaviour of concrete containing rubber.

6. Industrial Slag

Slag is a partially vitreous by-product of smelting ore to separate the metal fraction from the worthless fraction. Although they can be regarded as a mixture of metal oxides slags can, however, contain metal sulfides and metal atoms in the elemental form. Several waste materials are nowadays vitrified to produce slag too.

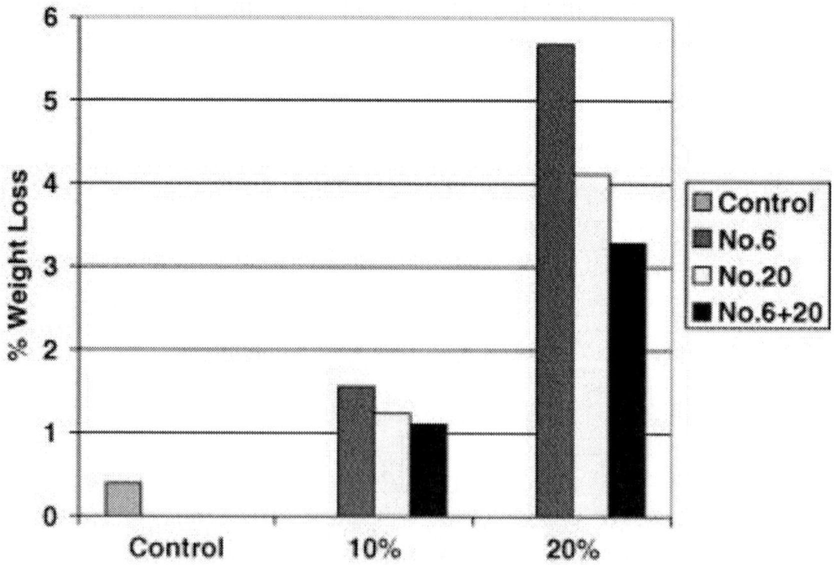

Figure 23. Abrasion resistance of concrete containing rubber tire crumb as partial substitution of aggregate at 10 and 20% ratio (No. 6 and No. 20 indicate rubber particles passing through ASTM 6 and 20 number sieves) [Sukontasukkul and Chaikaew 2006].

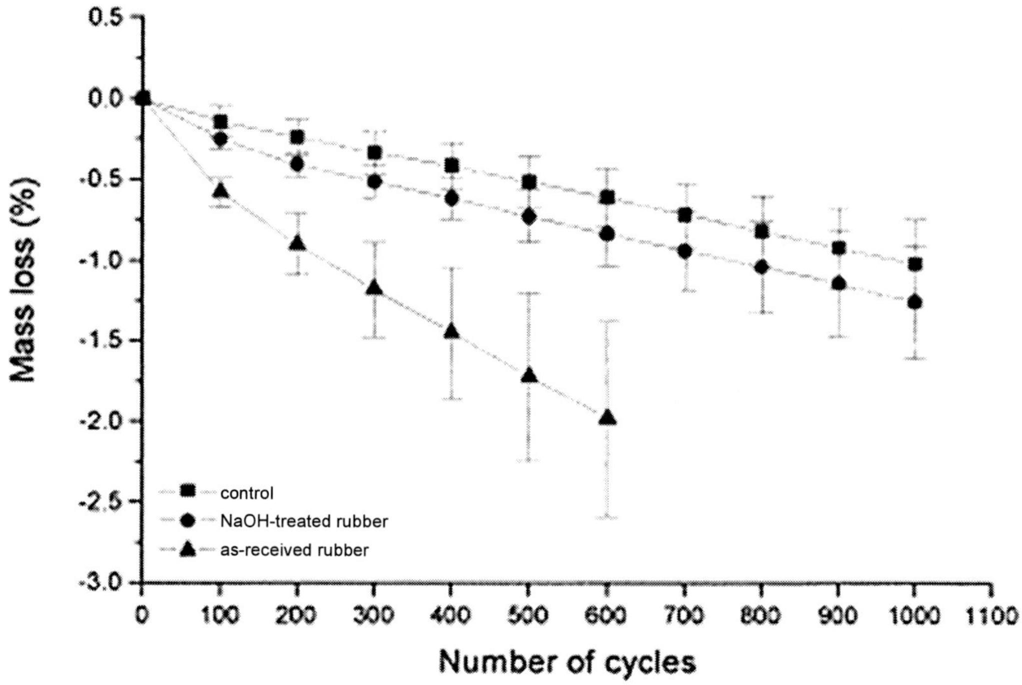

Figure 24. Mass loss from control cement paste and cement pastes containing 10% untreated and NaOH treated rubber powder [Segre and Joekes 2000].

6.1. Steel Slag

Steel slag is produced during the separation of the molten steel from impurities in steel-making furnaces. The slag occurs as a molten liquid melt and is a complex solution of silicates and oxides that solidifies upon cooling. There are several different types of steel slag produced during steel-making of which basic oxygen furnace steel slag (BOF slag), electric arc furnace slag (EAF slag) and ladle furnace slag (LDF slag) or refining slag are important.

An electric arc furnace produces steel by melting recycled steel scrap, using heat generated by an arc created by a large electric current. The slag is formed through the addition of lime, which is designed to remove impurities from within the steel. The slag has a lower density than steel and therefore floats on top of the bath of molten steel.

In the basic oxygen process, hot liquid blast furnace metal, scrap, and fluxes, which consist of lime (CaO) and dolomitic lime (CaO.MgO), are charged to a furnace. The oxygen is injected into the furnace where it combines with and removes the impurities in the charge. These impurities consist of carbon as gaseous carbon monoxide, and silicon, manganese, phosphorus and some iron as liquid oxides, which combine with lime and dolomitic lime to form the BOF steel slag.

After being tapped from the furnace, the molten steel is transferred in a ladle for further refining to remove additional impurities still contained within the steel. This operation is called ladle refining because it is completed within the transfer ladle. During ladle refining, additional steel slags are generated by again adding fluxes to the ladle to melt.

Steel mill scale is produced during processing of iron in steel mills. During the processing of steel in a steel mill, iron oxides, known as mill scale are formed on the surface of the metal during continuous casting, reheating, and hot rolling operations. The steel mill scale is removed by water sprays. The steel mill scale is somewhat similar to steel slag and therefore, like steel slag, it can be used in concrete production.

The use of steel slag in the preparation of blended cement, or cement clinker and as aggregate or hydraulic binder in road construction has been proposed by several authors for quite a long time [Mahieux et al. 2009, Monshi and Asgarani 1999, Shi and Qian 2000]. However, not much has been reported on the use of steel slag as aggregate in the preparation of cement mortar and concrete. Compared with the use of BOF slag and LDF slag as fine aggregate in concrete, there are more literature references on EAF slag.

Tables 5, 6 and 7 show the compositions of BOF slag, LDF slag and EAF-slag, respectively. Steel slag is basic in nature (ratio of concentrations of CaO + MgO to SiO_2 + Al_2O_3 greater than 1) [Tossavainen et al. 2007]. The order of this ratio for LDF slag, EAF slag and BOF slag can be arranged as EAF slag $\leq$ LDF slag $\leq$ BOF slag. The reactivity and free CaO content of steel slag increases with its increasing basicity (Shi and Qian, 2000). Compared with BOF slag and LDF slag, most EAF slags contain less CaO and MgO in their chemical compositions. The dominant minerals in EAF slag can be grouped into dicalcium silicate (C_2S), merwinite (C_3MS_2) or kirschsteinite (CFS), depending on the basicity of steel slag, of which kirschsteinite has the lowest basicity [Qian et al. 2002].

Even though the concentrations of some toxic metals in steel slag are higher than those in normal soil, the metals are strongly bound with the slag matrix and therefore are not readily leached. Therefore steel slag can not be considered as a "characteristically hazardous" material. [Proctor et al. 2000]. However, in a recent report higher leachability values (above the standard limit for inert landfill) of Cr and Mo from EAF slag were also reported [Tossavainen et al. 2007]. As the chemical composition of steel slag is highly variable its mineralogical composition also varies. Table 8 shows the mineralogical composition of various types of slag, as reported by various authors. Calcium silicates, solid solutions of CaO-FeO-MnO-MgO and free-CaO are the commonest minerals in steel slag.

Table 5. Chemical composition of BOF slags

Author	Lun et al. 2008	Xue et al. 2006	Mahieux et al. 2009	Tossavainen et al. 2007	Altun and Yilmaz 2002		Reddy et al. 2006
Country	PRChina	PRChina	France	Sweden	Turkey		India
Type	Steel slag (from BOF)	BOF-slag	Weathered BOF-slag	BOF-slag	BOF-slag		BOF-slag
					S1	S2	
SiO_2 (%)	10.10	13.71	11.8	11.1	18.01	18.87	15.3
Al_2O_3 (%)	1.70	3.80	2.00	1.90	2.61	2.91	1.30
Fe and Fe_xO_y (%)	Fe_2O_3=20.57	FeO= 21.85 Fe_2O_3= 3.24	Fe_2O_3=22.6	FeO= 10.9 Fe_2O_3= 10.7 Fe = 2.30	14.10	11.73	Fe (total) = 16.2
CaO (%)	40.65	45.41	47.5	45.0	37.02	37.94	52.3
MnO (%)	0.06	3.27	1.90	3.10	7.52	7.72	0.39
MgO (%)	10.88	6.25	6.30	9.60	14.10	14.37	1.1
Free lime (%)	nr	nr	nr	nr	0.83	0.83	10
Basicity: (CaO+MgO)/(Al_2O_3+SiO_2)	4.37	2.95	3.90	4.20	2.48	2.40	3.82

nr: not reported.

Table 6. Chemical composition of LDF-slags

Author	Setien et al. 2009			Tossavainen et al. 2007	Shi 2002	Rodriguez et al. 2009
Country	Spain			Sweden	PRChina	Spain
Type	1	2	3	LDF slag	LDF-slag fines	LDF-slag
SiO_2 (%)	15.0	19.8	12.6	14.2	26.4-26.9	17
Al_2O_3 (%)	12.5	4.3	18.6	22.9	4.3-5.2	11
Fe and Fe_xO_y (%)	Fe_2O_3=2.1	Fe_2O_3=3.3	Fe_2O_3=1.6	Fe_2O_3=1.1 FeO = 0.5 Fe = 0.4	Fe_2O_3=1.0-1.6	nr
CaO (%)	55.0	57.5	50.5	42.5	56.6-57.0	56
MnO (%)	0.36	0.42	0.52	0.2	0.5-1.0	
MgO (%)	7.5	11.6	11.9	12.6	3.2-4.2	10
Free lime (%)	19.0	3.5	9.5	nr	nr	nr
Free MgO (%)	3.0	10.0	8.0	nr	nr	nr
Basicity: (CaO+MgO)/ (Al_2O_3+SiO_2)	2.27	2.87	2.00	1.49	1.93	2.36

nr: not reported.

Table 7. Chemical composition of EAF-slags

Author	Qian et al. 2002	Luxan et al. 2000		Manso et al. 2006	Pellegrino and Gaddo 2009		Tossavainen et al. 2007	
Country	Singapore	Spain		Spain	Italy		Sweden	
Type	Kirschsteinite based EAF-slag	Black EAF-slag (fusion of scrap steel)		Black EAF-slag	EAF slag		EAF slag	
		S1	S2 (contaminated by white slag)		medium	coarse	S1	S2
SiO_2 (%)	19.38	6.04	15.35	15.3	10.1	14.7	32.2	14.1
Al_2O_3 (%)	5.27	14.07	12.21	7.4	5.70	7.20	3.70	6.70
Fe and/or Fe_xO_y (%)	FeO = 30.54 Fe_2O_3= 8.49	FeO= 27.41	FeO= 34.36	FeO+Fe_2O_3=42.5	FeO= 37.2	FeO= 44.8	FeO= 3.30 Fe_2O_3= 1.00 Fe = 0.10	FeO= 5.60 Fe_2O_3= 20.3 Fe = 0.60
CaO (%)	22.48	29.11	24.40	23.9	24.2	29.5	45.5	38.8
MnO (%)	1.10	15.58	5.57	4.5	5.10	5.70	2.0	5.0
MgO (%)	9.51	3.35	2.91	5.1	1.90	4.60	5.2	3.9
Free lime (%)	nd	nr	nr	0.45	nr	nr	nr	nr
Free MgO (%)	nr	nr	nr	~1.0	nr	nr	nr	nr
Basicity: (CaO+MgO)/(Al_2O_3 +SiO_2)	1.30	1.61	0.99	1.28	1.65	1.60	1.41	2.05

nd: not detected; nr: not reported.

Table 8. Mineralogical composition of various slags

Authors	Country	Type of slag	Minerals identified
Rojas and Rojas 2004	Spain	EAF-slag	Wustite/Plustite, Magnesioferrite/Magnetite, Hematite, Larnite, Bredigite/Merwinite, Gehlenite, Birnessite/Groutellite (Manganese oxides)
Pellegrino et al. 2007	Italy	EAF	Fine Fraction: Glaucochroite $(CaMn)_2SiO_4$, Iron manganese oxide (FeMnO)
			Medium: Calcite ($CaCO_3$), Glaucochroite $(CaMn)_2SiO_4$), Iron manganese oxide (FeMnO), Magnetite (Fe_3O_4), Portlandite ($Ca(OH)_2$)
Tossavainen et al. 2007	Sweden	LDF-slag	Mayenite, ($Ca_{12}Al_{14}O_{33}$), free MgO. β-Ca_2SiO_4, γ-Ca_2SiO_4, $Ca_2Al_2SiO_7$
		BOF-slag	Larnite, β-Ca_2SiO_4, solid solution of (Fe,Mg,Mn)O, solid solution of (CaMg)O
		EAF-slag, S1	Merwinite, $Ca_3Mg(SiO_4)_2$, γ-Ca_2SiO_4, solid solution spinel phase (Mg, Mn)(Cr, Al)$_2$O$_4$
		EAF-slag, S2	β-Ca_2SiO_4, wustite-type solid solution ((Fe, Mg, Mn)O), $Ca_2(Al, Fe)_2O_5$, Fe_2O_3
Qian et al. 2002	Singapore	Kirschsteinite based EAF-slag	Kirschsteinite ($CaFe^{2+}(SiO_4)$), Mg-wustite
Luxan et al. 2000	Spain	EAF-slag, S1 and S2	Major: Gehlenite [Ca_2 $Al(Al,Si)_2$ O_7], Larnite (Ca_2SiO_4) Minor: Bredigite [$Ca_{14}Mg_2(SiO_4)_8$], Manganese oxides (Mn_3O_4, MnO_2), Magnesioferrite ($MgFe_2O_4$), Magnetite (Fe_3O_4)

As steel slag contains expansive materials like CaO and MgO, slag is generally treated if it is to be used in concrete. Slag may be aged or weathered, or treated by steam and autoclave curing to reduce the expansive oxide contents [Chen et al. 2007, Faraone et al. 2009, Lun et al. 2008, Pellegrino and Gaddo 2009]. The slump, bulk density, and unit weight of fresh concrete and the different strength properties, elasticity modulus, and drying shrinkage of hardened concrete are normally evaluated. Slag should be properly graded to obtain better performance of concrete containing steel slag as aggregate [Manso et al. 2006]. As most steel slags contain free CaO and MgO, experiments are generally performed to evaluate the free CaO and MgO content in the slag as well as the expansion behaviour (i.e. internal damage) of concrete and mortar containing steel slag. Some other durability tests like drying shrinkage, alkali-aggregate reaction, freezing-thawing, and wetting-drying cycles are also performed. Leaching of toxic elements from concrete containing slag is also reported by some authors, because of the presence of toxic elements in steel slag.

As the chemical composition of one steel slag is different from that of others, proper comparison of the properties of two steel slags is not possible. There are some slags [Maslehuddin et al. 2003, Qasrawi et al. 2009] which contain more than 85% of Fe_2O_3 and these are therefore completely different from other EAF slags (Table 9). The properties of concrete containing these slags cannot be compared with those of concrete containing other slags. The behaviour of concrete containing these slags and steel mill scale [Al-Otaibi 2008] is discussed first. The properties of other steel slags are discussed according to their basicity (ratio of (CaO + MgO) to (Al_2O_3+SiO_2)). As stated earlier, the reactivity and free CaO content of steel slag increases with its increasing basicity [Shi and Qian, 2000].

Table 9. Chemical composition of some steel slag described as steel slag or steel making slag and steel mill scale

Author	Faraone et al. 2009	Al-Negheimish 1997	Chen et al. 2007	Maslehuddinn et al. 2003	Quasrawi et al. 2009	Al-Otaibi 2008
Country	Italy	Saudi Arabia	PR China	Saudi Arabia	Jordan	Kuwait
Type	Steel slag	Steel making slag (EAF-slag)	Steel slag	Steel slag	Unprocessed steel slag	Steel mill scale
SiO_2 (%)	13.2	17.43	13.40	1.00	0.80	1.37
Al_2O_3 (%)	6.80	7.85	1.04	nr	nr	0.10
Fe and/or Fe_xO_y (%)	Fe_xO_y = 9.80	Total Fe = 20.09 FeO = 18.33	FeO = 18.98 Fe_xO_y =8.45	Fe_xO_y =89.0	Fe_xO_y= 97.05	Fe_xO_y= 94.61
CaO (%)	37.9	30.53	51.51	$CaCO_3$= 10	0.40	0.11
MnO (%)	13.5	2.66	1.65		1.07	1.03
MgO (%)	11.9	13.19	3.45		0.40	0.03
Basicity: (CaO+MgO)/(Al_2O_3+SiO_2)	2.49	1.73	3.81	-	-	-

nr: not reported.

Table 10. Some properties of concrete specimens prepared using crushed limestone and varying percent of steel slag aggregate [Masleduddin et al. 2003]

Property	Crushed limestone aggregate (60%)	Steel slag aggregate				
		45%	50%	55%	60%	65%
Unit weight, kg/m^3	2330	2436	2523	2653	2736	2769
Compressive strength*, MPa	39.7	31.4	37.7	37.6	41.6	42.7
Flexural strength*, MPa	3.96	3.47	3.61	3.69	3.81	4.21
Splitting tensile strength*, MPa	6.33	5.79	6.06	6.11	6.26	6.35
Water absorption*, %	5.33	4.51	4.33	3.89	3.71	3.64
Volume permeable voids*, %	13.6	13.58	11.87	11.69	11.12	10.04
Pulse velocity*, km/s	4.55	4.60	4.62	4.64	4.78	4.96

* Properties were determined after 28 days of normal curing.

6.1.1. Steel Slag/steel Mill Scale with very High Fe_2O_3 Content (≥89%)

The incorporation of EAF slag (Fe_2O_3 > 97%) as fine aggregate in concrete reduces slump of concrete owing to the presence of large amount of fine particles in the slag and the angular nature of slag particles [Qasrawi et al. 2009]. The unit weight (Table 10) and density of concrete containing steel slag (Fe_2O_3 ≥ 89%) as either coarse or fine aggregate are greater than those of conventional concrete because steel slag has a higher specific gravity than the conventional aggregate used in concrete preparation [Masleduddin et al. 2003, Qasrawi et al. 2009].

Concrete with EAF slag (Fe_2O_3 > 97%) as fine aggregate up to 30% replacement of natural fine aggregate exhibits higher compressive strength up to 180 days of curing [Qasrawi et al. 2009]. Increasing the replacement ratio of fine aggregate by steel mill scale (Fe_2O_3 > 94%) increases the compressive strength of the resulting concrete and replacement ratio of 40% gives the highest compressive strength [Al-Otaibi 2008]. The tensile strength of concrete with the same EAF slag (Fe_2O_3 > 97%) as fine aggregate up to 100% is better than that of conventional concrete due to the higher angularity of EAF slag compared with natural fine aggregate, and optimum strength occurs when the sand is replaced by 30-50% of EAF slag [Qasrawi et al. 2009]. The elimination of materials finer than 0.15 mm from EAF slag (Fe_2O_3 > 97%) can improve the compressive strength of the resulting concrete that contains EAF slag as fine aggregate [Qasrawi et al. 2009]. The compressive, flexural and splitting tensile strengths and modulus of elasticity of the concrete with steel slag as coarse aggregate are better than those of conventional concrete (Table 10) [Maslehuddin et al. 2003].

The pulse velocity in the concrete containing steel slag aggregate (Fe_2O_3 = 89%) as coarse aggregate is higher than that of concrete containing crushed limestone aggregate, which indicates that the incorporation of steel slag in concrete produces a denser concrete than that which contains limestone aggregate (Table 10) [Maslehuddin et al. 2003]. The water absorption capacity and the volume of permeable voids of concrete containing steel slag aggregate (Fe_2O_3 = 89%) as coarse aggregate are lower than in concrete containing limestone aggregate, and these values decrease the greater the content of steel slag in the concrete mix (Table 10) [Maslehuddin et al. 2003]. Concrete containing steel slag aggregate (Fe_2O_3 = 89%) as coarse aggregate performs better than that which contains crushed limestone aggregate when both mixes are exposed to similar thermal cycles [Maslehuddin et al. 2003]. The drying shrinkage of cement mortar containing steel mill scrape (Fe_2O_3 > 94%) and steel slag aggregate (Fe_2O_3 = 89%) is lower than that of conventional cement mortar [Al-Otaibi 2008, Maslehuddin et al. 2003]. Cement mortar containing steel slag (Fe_2O_3 = 89%) as coarse aggregate expands slightly in moist curing, albeit within the range of standard specifications [Maslehuddin et al. 2003]. Concrete with steel slag (Fe_2O_3 = 89%) aggregate resists reinforcement corrosion better than that containing limestone aggregate [Maslehuddin et al. 2003].

6.1.2. EAF, BOF and LDF Steel Slag with Fe_2O_3 Content <50%

6.1.2.1. EAF Slag

Concrete made with EAF slag (basicity 1.73, 1.29, 1.59) as aggregate exhibits almost similar (or even better) compressive, flexural and tensile strengths and Young's modulus than those observed for conventional concrete in normal curing conditions [Al-Negheimish 1997,

Manso et al. 2006, Pellegrino and Gaddo 2009]. Concrete containing EAF slag (basicity 1.73) as coarse aggregate exhibits higher stiffness than concrete containing natural gravel [Al-Negheimish 1997]. The compressive strength of cement mortar containing steel-making slag (basicity 2.49) as aggregate can be improved if the slag contains a large amount of coarse particles [Faraone et al. 2009]. Since the fine fraction of steel-making slag (basicity 2.49) affects the water to cement ratio, it is necessary to consider the water to (cement + fine fraction of slag) ratio during the preparation of cement mortar containing this steel slag as aggregate [Faraone et al. 2009].

The drying shrinkage of concrete with EAF slag (basicity 1.73) is lower than that of conventional concrete with natural aggregate [Al-Negheimish 1997]. The compressive strength of concrete with EAF slag (basicity 1.29, 1.59) as aggregate is lower than that of concrete with traditional aggregate when both concretes are exposed to similar curing conditions (autoclave curing, hot water curing, hot water curing followed by accelerated aging, repeated freezing and thawing, repeated wetting and drying) due to the presence of free CaO and MgO in the EAF slag and the porosity in the slag-cement interface. However, the compressive strength of concrete with EAF slag cured in hot water followed by accelerated aging is better than that of the same concrete cured in hot water [Manso et al. 2006, Pellegrino and Gaddo 2009]. Concrete with EAF slag as aggregate expands slightly owing to alkali-aggregate reaction (measured according to ASTM C-1250), which may be due the presence of free lime and periclase in EAF slag; however, this expansion is lower than the critical value [Manso et al 2006]. The use of EAF slag as fine aggregate in concrete causes pop-outs due to the presence of free CaO and free MgO in the EAF slag [Lee and Lee 2009], as seen in Figure 25.

6.1.2.2. BOF Slag

The hydration of over-burnt free lime and periclase phases in BOF slag (basicity 4.37) is the main reason for cracking in cement mortar containing this type of slag as aggregate [Lun et al. 2008]. The volume stability of BOF slag (basicity 4.37) can be improved significantly by steam and autoclave treatment. However, autoclave treatment of slag is better than steam curing due to the different hydraulic activity of free lime and periclase [Lun et al. 2008]. The use of steel slag (basicity 3.81) instead of sand in cement mortar containing ground granulated blast furnace slag, cement clinker and gypsum lowers the amount of clinker content in mortar and superior strength properties are obtained for mortar containing 10% clinker [Chen et al. 2007].

6.1.2.3. LDF Slag

Reports on the use of LDF slag as aggregate in the preparation of cement mortar and concrete are few and far between. But there is a report on the use of LDF slag as partial substitution of sand and cement in the preparation of masonry cement mortar. Incorporating LDF slag with cement and sand increases the density and water demand and decreases the amount of entrained air of the resulting fresh and masonry mortar. LDF slag incorporation with cement and sand also improves the workability of masonry mortar mixes. The evaluation of the mechanical performance of masonry mortar containing LDF slag demonstrates its weak hydraulic activity at the early stage of curing and its considerable hydraulic activity at the latter stage of curing [Rodriguez et al. 2009].

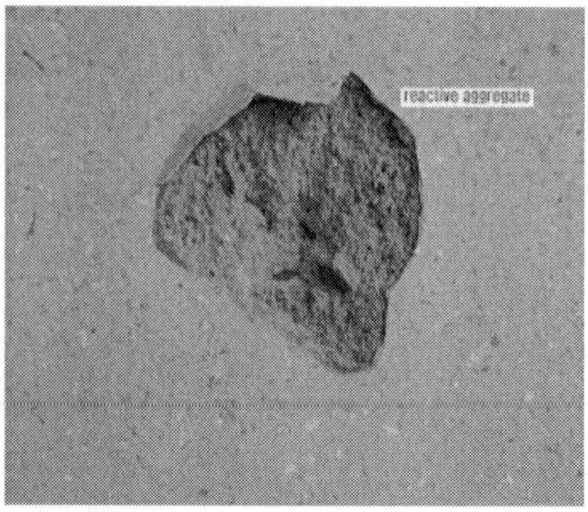

(A)

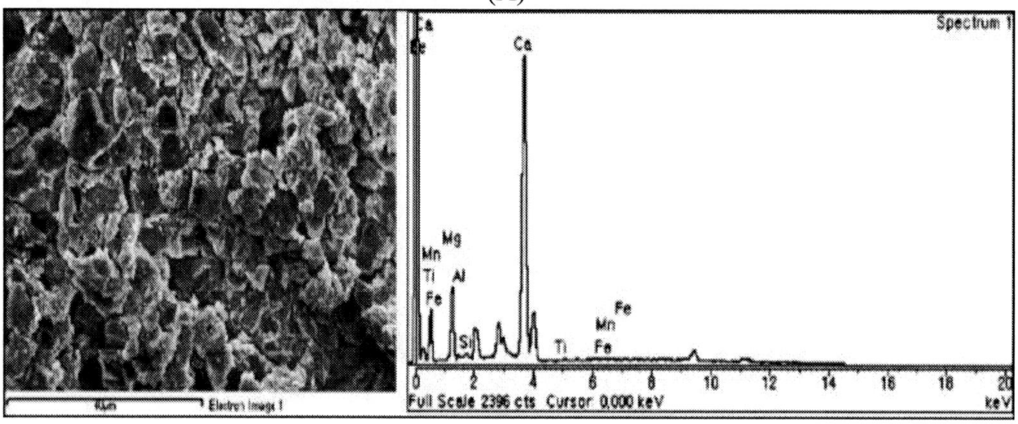

Specimen	Major chemical composition (wt.%)							
	Ca	**Si**	**Al**	**Le**	**S**	**K**	**Mg**	**Cr**
A	75.7	4.35	1.68	1.18	–	–	10.5	–
B	77.9	4.46	1.83	–	–	–	9.73	–
C	67.8	9.09	8.96	–	–	–	2.29	6.49

(B)

Figure 25. (A) Example of pop-out on concrete surface highlighting the reactive substance; (B) a specimen SEM micrograph and EDX spectrum of the reactive substance plus the elemental composition of some EDXA results [Lee and Lee 2009].

In comparison with BOF and LDF slag, more references are available on the use of EAF slag as aggregate in concrete. But considerable work on using these types of slag is needed because they contain deleterious components, mostly free lime and magnesia.

6.2. Non-ferrous Industries Slag

Non-ferrous metallurgical slags (called from here on non-ferrous slag) are generated during refining of metals such as Cu, Cr, Zn and the treatment of waste such as Pb-acid batteries. Several references are available on the use of copper slag as aggregate in cement mortar and concrete, including a review. In this section, the performance of copper slag as aggregate is discussed.

Copper slag is a by-product obtained during matte (molten copper sulfide) smelting and copper refining [Biswas and Davenport 2002]. Major constituents of a smelting charge are sulphides and oxides of iron and copper. The charge also contains the oxides Al_2O_3, CaO, MgO, and principally SiO_2, which are either present in the original concentrate or added as flux. As a result, copper-rich matte (sulphides) and copper slag (oxides) are formed as two separate liquid phases. The molten slag is discharged from the furnace at 1000–1300 °C. When liquid slag is cooled slowly it forms a dense, hard crystalline product where a quick solidification by pouring molten slag into water yields an amorphous granulated slag. Production of one ton of copper generates approximately 2.2–3 tons of copper slag. In the United States about four million tons of copper slag are produced annually [Collins and Ciesielski 1994], in Japan it is about two million tons per year [Ayano and Sakata 2000], and approximately 360,000, 244,000 and 60,000 tons of copper slag are produced every year in Iran, Brazil and Oman, respectively [Behnood 2005, Moura et al. 1999, Taeb and Faghihi 2002].

Recycling, recovery of metal, production of value added products and disposal in slag dumps or stockpiles are the options for the management of copper slag. It has been widely used for abrasive tools, roofing granules, cutting tools, abrasive, tiles, glass, road-base construction, railroad ballast, asphalt pavements, cement clinker and blended cement production [Shi et al. 2008]. Many researchers have investigated the use of copper slag as fine or coarse aggregate in the preparation of cement mortar and concrete.

The main concern in the large-scale use of copper slags, including in construction, is the content of toxic elements in copper slag and the consequent leaching from slag and slag based products. But copper slag has been excluded from the listed hazardous waste category of the United States Environmental Protection Agency (USEPA). The United Nations (UN) Basel Convention on the Transboundary Movement of Hazardous Waste and their Disposal also ruled that copper slag is not a hazardous waste [Alter 2005].

The chemical composition of copper slag depends on the type of furnace, the metallurgical production process, and the composition of the extracted ore. The percentage range of the main oxides of copper slag can vary as follows - Fe_2O_3: 35–60%, SiO_2: 25–40%, CaO: 2–10%, Al_2O_3: 3–15%, CuO: 0.3–2.1%, MgO: 0.7–3.5%. The density of copper slag varies between 3.16 and 3.87 g/cm^3, the average specific gravity of copper slag is about 3.5 g/cm^3 and the average water absorption of copper slag is 0.15-0.55. Water-cooled copper slag has a higher water absorption and lower unit weight than air-cooled copper slag owing to its more porous texture [Shi et al. 2008].

6.2.1. Properties of Copper Slag as Aggregate in Conventional Concrete

The effect of copper slag on the setting time depends on its particle size since the smaller the particle size the longer the delay. Washing can remove the 0.15 mm size fraction of copper slag, however, which improves the setting time of concrete containing copper slag as aggregate [Ayano and Sakata 2000, Shoya et al. 1997]. There was more bleeding of concrete as the copper slag incorporation increased, and this was due to the high specific gravity and glass-like surface properties with irregular grain shape [Shoya et al. 1997]. Bleeding is dependent on the water to cement ratio, the volume fraction of slag, and the air content. 40% copper slag can be used as partial replacement of aggregate to control the amount of bleeding to less than 5 l/m^2 [Shoya et al. 1997]. The addition of a chemical admixture compound with cellulose ether and a powder-like material, such as lime stone powder, is highly effective at improving the bleeding property of concrete containing copper slag as aggregate [Shoya et al. 1997]. One study, however, reported comparatively less bleeding of mortar containing copper slag as fine aggregate than occurred in that containing natural sand [Hwang and Laiw, 1989].

The compressive and tensile strength of concrete specimens made with copper slag fine aggregate are almost the same as for conventional concrete or even significantly more than control mixes [Ayano and Sakata 2000, Caliskan and Behnood 2004, Hwang and Laiw 1989, Li 1999, Shoya et al. 1997, Zong 2003], as seen in Figure 26.

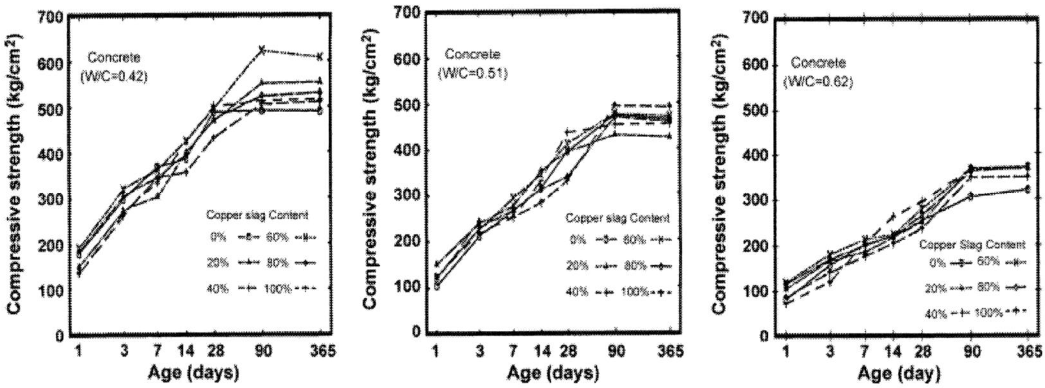

Figure 26. Compressive strength of concrete containing copper slag as fine aggregate at different water to cement ratios [Hwang and Laiw 1989].

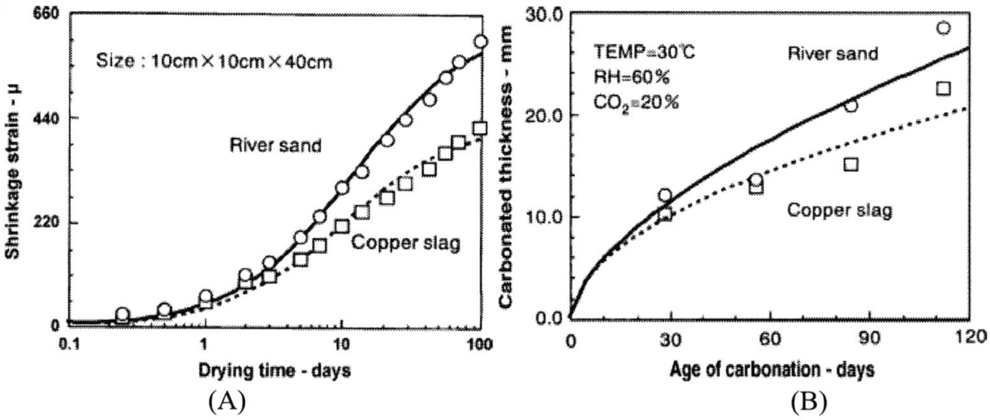

Figure 27. A comparison of (A) drying shrinkage strain and (B) carbonation depth of concrete containing copper slag as fine aggregate with conventional concrete containing river sand [Ayano and Sakata 2000].

The drying shrinkage of concrete containing copper slag as fine aggregate is similar to or even less than that of specimens without copper slag [Ayano and Sakata 2000], as seen in Figure 27(A). The abrasion resistance of the cement mortar containing copper slag aggregate is better than for conventional concrete [Tang et al. 2000]. The resistance to sulfate attack and the rate of carbonation of concrete containing copper slag aggregate are very similar to (or even better than) the ones of concrete containing conventional aggregate [Ayano and Sakata

2000, Hwang and Laiw 1989], as seen in Figure 27(B). The freezing–thawing resistance of concrete containing copper slag aggregate is lower than that of conventional concrete owing to the internal defects originated by the up flow of bleeding water. But the addition of chemical admixture and lime stone powder improves this property [Shoya et al. 1997].

6.2.2. Properties of Copper Slag as Aggregate in High-strength Concrete (HSC)

The slump (and therefore the workability) of high strength concrete (HSC) containing copper slag as fine aggregate is higher than that of conventional concrete and it increases significantly with the increase of copper slag content in concrete mixes, due to the low water absorption characteristics of copper slag and its glassy surface compared with that of sand, which causes surplus quantity of free water in the concrete mix. The density of HPC containing copper slag as fine aggregate is higher than that of conventional concrete because of the higher specific gravity of copper slag compared with sand [Al-Jabri et al. 2009].

The 7- and 28-day compressive, splitting tensile and flexural strength of HSC containing copper slag up to 50% replacement ratio (of natural sand) as fine aggregate are almost the same as or slightly better than those of HSC containing natural sand. Beyond that level compressive strength is reduced significantly owing to the marked increase in free water which causes the particles of the constituents to separate, leaving pores in the hardened concrete (Figure 28) [Al-Jabri et al. 2009]. The compressive and splitting tensile strength of HSC containing copper slag as coarse aggregate and prepared at constant water to cement ratio is higher than for HSC containing limestone aggregate; this is because of the higher strength of copper slag aggregate compared with limestone aggregate and the porous and rough surface texture of copper slag. This surface texture may produce a superior bond and transition zone than the limestone aggregate [Khanzadi and Behnood 2009]. Figure 29 shows the surface of concrete containing limestone and copper slag as coarse aggregate after determination of compressive strength. An extended crack is propagated along the interface between cement paste and limestone aggregate (a), but no such cracks are observed on the surface of concrete containing copper slag (b). The incorporation of silica fume with cement can produce a stronger transition zone between the copper slag aggregate and the cement paste because of its pozzolanic reaction [Khanzadi and Behnood 2009].

Surface water absorption decreases with increasing copper slag content in the HSC, up to 40% of substitution; after this, water absorption rapidly increases [Al-Jabri et al. 2009]. HSC concrete containing copper slag as coarse aggregate exhibits higher rebound hammer value than one containing limestone aggregate, which is due to the higher strength of copper slag aggregate compared with limestone aggregate [Khanzadi and Behnood 2009].

6.2.3. Other Non-ferrous Slags

The use of other non-ferrous industrial slags as aggregate in concrete and cement mortar is also reported. The lead slag generated from lead smelting and from recycling of secondary lead batteries suggest that this slag can be used as fine and coarse aggregate in the preparation of cement mortar and concrete. Concrete containing some lead slag as aggregate shows better mechanical properties than conventional concrete [Atzeni et al. 1996, Metwally et al. 2005, Penpolcharoen 2005, Saikia et al. 2008]. The concentrations of the toxic elements present in the leachate generated from lead slag-containing cement mortar and concrete meet the standard specifications [Atzeni et al. 1996, Penpolcharoen 2005, Saikia et al. 2008].

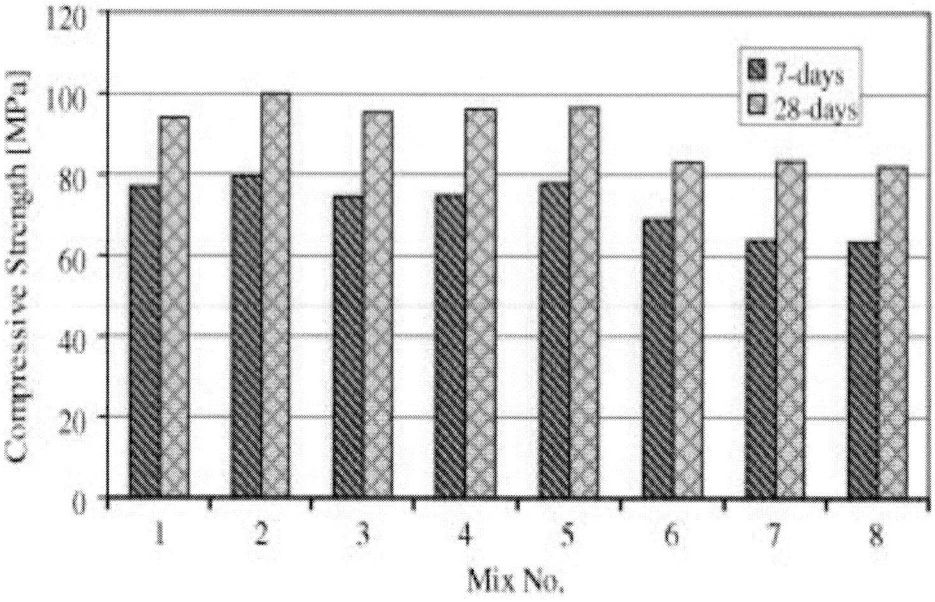

Figure 28. 7 and 28 days compressive strength of high-strength concrete containing copper slag as aggregate (mixes No. 1, 2, 3, 4, 5, 6, 7 and 8 represent 0, 10, 20, 40, 50, 60, 80 and 100 % replacement of fine aggregate by copper slag, respectively) [Al-Jabri et al. 2009].

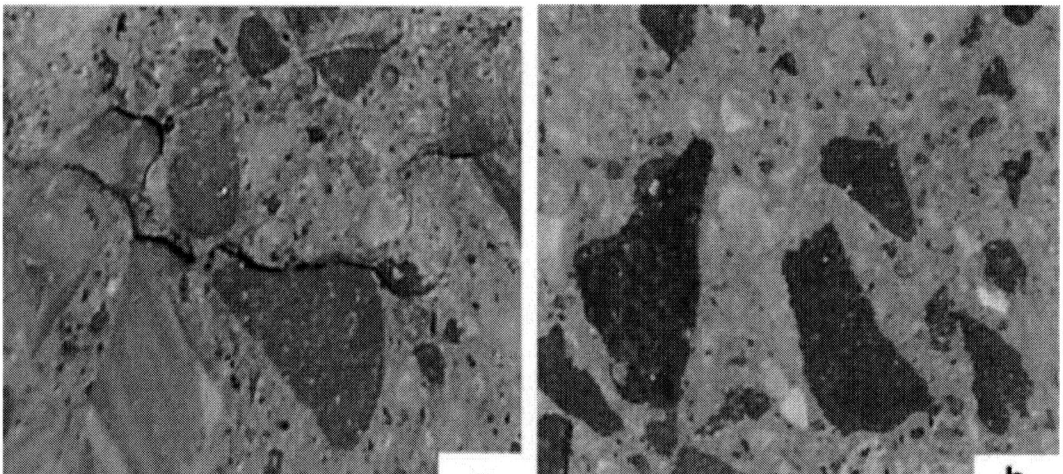

Figure 29. Comparison of the surfaces of failure specimens containing: (a) limestone and (b) copper slag [Khanzadi and Behnood 2009].

Ferro-silicate slag, generated during the production of zinc in an Imperial Smelting Furnace (ISF slag) has successfully been used as aggregate in concrete road construction [Morrison et al. 2003, Morrison and Richardson 2004]. The use of other slags, such as ferronickel slag, ferrochromium slag, slag generated during processing of electric arc furnace dust and residues recovered in zinc processing industry (Waelz slag), aluminium containing salt slag, zinc slag as fine or coarse aggregate, is also reported [Atzeni et al. 1996, Monosi et al. 2001, Pereira et al. 2000, Shoya et al.

1997, Sorlini et al. 2004, Zelic 2005]. But thorough investigations into the environmental suitability as well as the long term mechanical performance and durability of concrete containing slag are necessary if these materials are to be used as aggregate in concrete.

7. Other Waste as Aggregate, Filler or Fibre in Concrete

The literature contains many references on the use of numerous other waste materials as aggregate, filler or fibre in concrete. The behaviour of some waste materials, which can be considered as industrial by-products, is discussed here.

7.1. Waste from Food Industries

A preliminary study on the evaluation of compressive strength and leachability of toxic elements from cement mortar containing low risk meat and bone meal bottom ash (MBMBA) as fine aggregate indicates its potential for use as partial substitution of the sand fraction (up to 30% level) in the preparation of cement mortar [Cyr and Ludmann 2006]. The content of chloride and Cr, Cu, Ni, Zn, in MBMBA must be considered, however, before any such application. Figure 30 shows the XRD pattern of MBMBA and ratio of the strength development of cement mortar containing a different percentage of MBMBA in partial substitution of fine aggregate.

Lightweight concrete containing tobacco waste as fine aggregate, collected from a cigarette factory can be considered as lightweight concrete class according to values of consistency, unit weight (0.50-0.56 kg/dm^3, porosity, compactness, compressive strength (0.20-0.60 N/mm^2) and thermal conductivity 0.194-0.210 W/mK[13]). This type of concrete can be used as a coating and dividing material in buildings owing to its thermal insulating behaviour [Ozturk and Bayrak 2004].

Disposal of spent mushroom substrate (SMS) has been a major problem facing farmers, especially in areas where yield far exceeds demand. Concrete containing quicklime-treated SMS as partial substitution of sand can be used for sidewalks, concrete kerbs, concrete barricades, sound walls, and other non-structural applications [Pang et al. 2007]. But the durability of this kind of concrete needs further investigation.

Yang et al. [2005, 2009] studied the effect of crushed oyster shell (OS) as a partial substitution of sand at 10% and 20% in concrete. From the experimental results they concluded that:

- The workability of concrete containing OS decreased with increasing fineness modulus of OS and with increasing substitution of sand by OS; setting time was not affected by partial substitution of sand by OS;
- The 1-year compressive strength of concrete containing OS at the 10% and 20% level was lower than the conventional concrete (Figure 31(A)); the elastic modulus of concrete containing OS at all replacement levels was lower than that of conventional concrete up to 1 year of curing;

- The drying shrinkage of concrete containing OS at all replacement levels was higher than that of conventional concrete (Figure 31(B)); incorporation of OS as partial replacement of sand in concrete improved the freezing and thawing resistance of conventional concrete;
- The permeability of concrete containing OS was lower than that of conventional concrete; on the other hand, the partial substitution of OS as fine aggregate in concrete did not affect the resistance to carbonation and acid attack compared to that of conventional concrete.

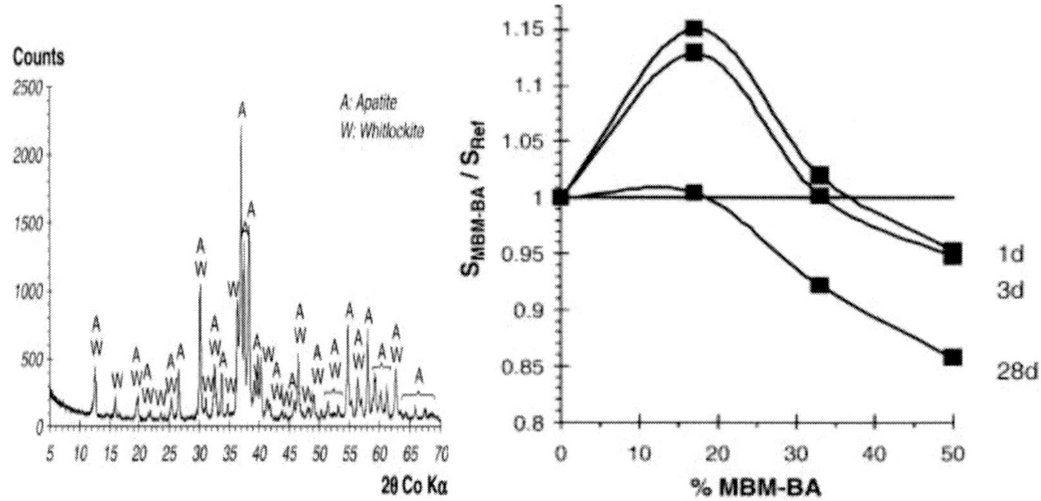

(A) (B)

Figure 30. (A) XRD pattern of meat and bone meal bottom ash; (B) Ratio of strength at 1, 3 and 28 days of mortars containing 17%, 33% and 50% of MBM-BA as replacement of sand, compared with reference mortar (0% MBM-BA) mortar [Cyr and Ludmann 2006].

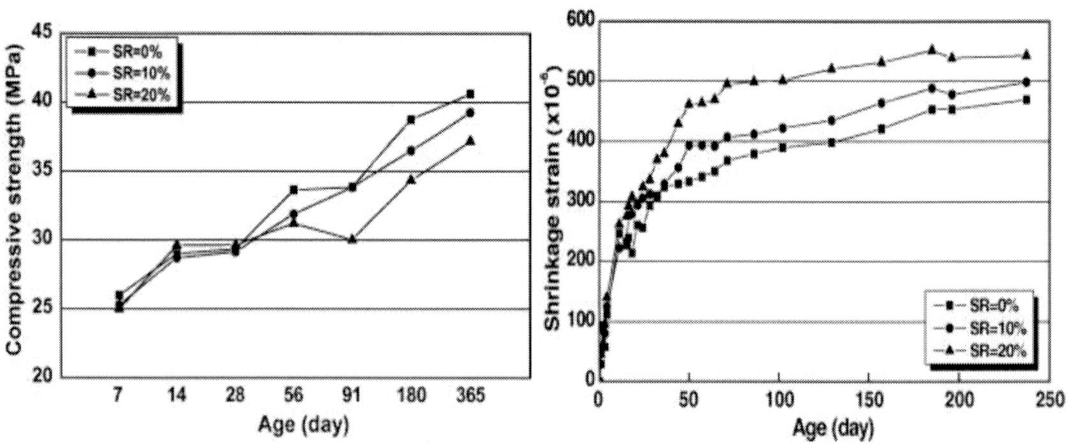

(A) (B)

Figure 31. Properties of concrete containing crushed oyster shell as partial substitution of fine aggregate: (A) Compressive strength; (B) Drying shrinkage [Yang et al. 2009].

A detailed study is available on the use of oil palm shell (OPS), an agricultural waste produced in palm oil producing countries, as coarse lightweight aggregate in structural lightweight concrete production. OPS can be used as coarse aggregate for the manufacture of structural lightweight concrete with comparable concrete properties, including durability. Proper curing is essential for OPS concrete to achieve better performance at later ages. Treatment with polyvinyl alcohol can improve the aggregate properties of OPS [Basri et al. 1999, Jumaat et al., 2009, Mannan et al. 2006, Mannan and Ganapathy 2001a, 2001b, 2002, 2004, Teo et al. 2006, 2007, 2010].

7.2. Pulp and Paper Mill Waste

A few studies are available on the use of waste generated from pulp and paper mills as fine aggregate, filler or fibre in concrete [Ahmadi and Al-Khaja 2001, Gallardo and Adajar, 2006, Naik 2002, Naik et al. 2004]. The use of dry paper mill sludge as fine aggregate in concrete up to 10% replacement of fine aggregate can improve the compressive and splitting tensile strength of resulting concrete [Gallardo and Adajar, 2006]. Ahmadi and Al-Khaja reported that the incorporation of paper sludge as aggregate decreased the compressive and splitting tensile strength of resulting concrete. Concrete masonry blocks containing sludge at 5% replacement level exhibit 8 MPa compressive strength, 1.3 MPa splitting strength, 11.9% water absorption, and 20 kN/m^3 density [Ahmadi and Al-Khaja 2001].

Fibres and residue from pulp and paper mill waste as well as de-inking solids from paper-recycling plants can be used in concrete. However, these materials should be properly dispersed in water, preferably hot water, before such sludge is used in structural-grade Portland cement concrete. Chloride ion penetration can be reduced by adding residual solids to concrete. This type of concrete also showed higher resistance to salt scaling and freezing-and thawing damage than control concrete [Naik 2002, Naik et al. 2004].

7.3. Leather Waste

Baffa and Akasaki [2005] investigated the performance of lightweight concrete prepared with leather waste, mostly residue from cattle, as an aggregate. Results showed that hardened mortar specimens that contain bits of leather with dimensions larger than 10 mm can be permanently deformed if the mortar is cured in humid condition. The compressive strength, the splitting tensile strength and drying shrinkage are within the standards' limits but they may differ for different types of leather. The expansion of concrete that contains leather is lower than that of the control concrete and it decreases as the amount of leather pieces in concrete increases [Baffa and Akasaki 2005].

7.4. Different Sludge

Rao et al. [2009] reported that the partial replacement of river-sand by bone char sludge, a waste material generated during the purification of fluoride contaminated water, did not degrade the strength or environmental integrity of concrete specimens.

Kuo et al. [2007] studied the use of organo-modified reservoir sludge (OMRS) as fine aggregate in cement mortars. They concluded that OMRS can be used to replace 30% of fine aggregate for the preparation of normal-strength cement mortars with improved water permeability (Figure 32). The OMRS particles can be used in controlled low-strength materials if their replacement percentage for fine aggregates is higher than 80%.

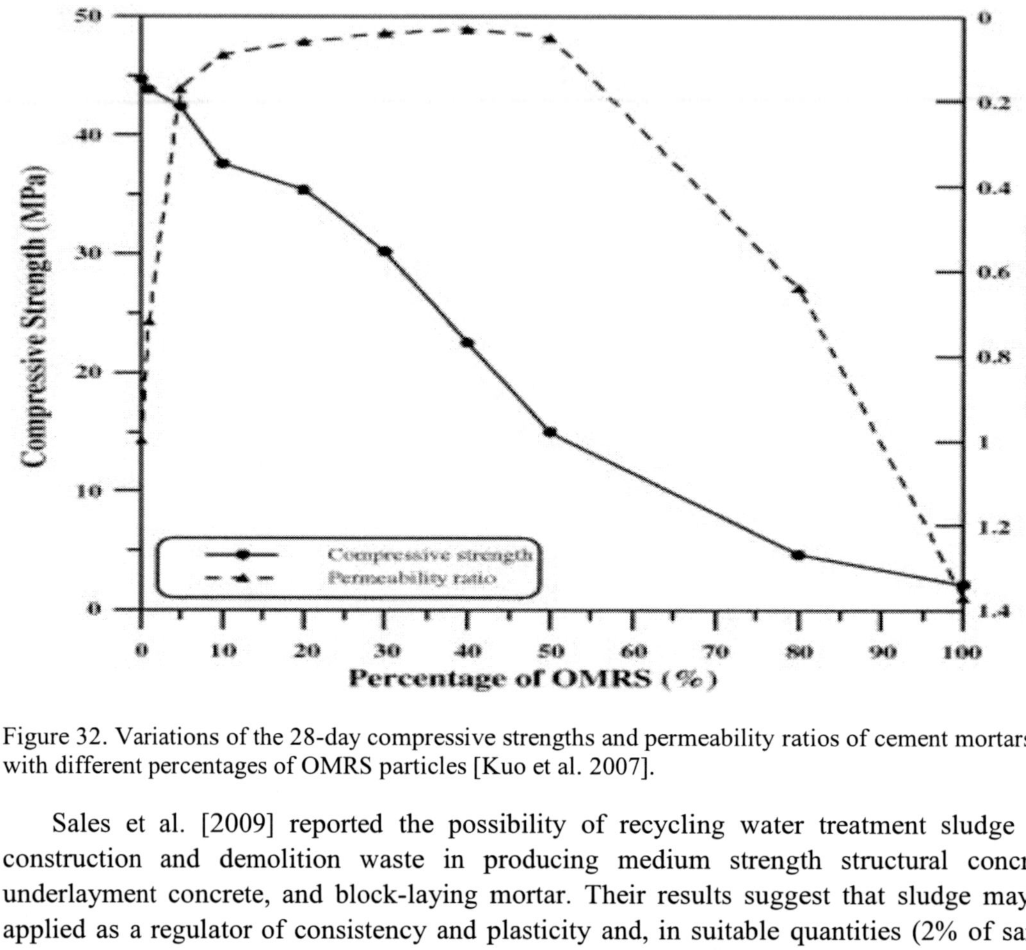

Figure 32. Variations of the 28-day compressive strengths and permeability ratios of cement mortars with different percentages of OMRS particles [Kuo et al. 2007].

Sales et al. [2009] reported the possibility of recycling water treatment sludge and construction and demolition waste in producing medium strength structural concrete, underlayment concrete, and block-laying mortar. Their results suggest that sludge may be applied as a regulator of consistency and plasticity and, in suitable quantities (2% of sand), can even increase the compressive strength of concrete and mortar produced for a given application.

Concrete containing municipal water treatment plant sludge as partial replacement of fine aggregate up to 30% level meets the ACI requirements of compressive and flexural strength of normal weight concrete. Moreover the skid resistance of concrete pavement blocks prepared by replacing 30% of fine aggregate by sludge ash is higher than that of normal pavement blocks [Khanbilvardi and Afshari 1999].

Aspiras and Manalo [1995] produced a concrete-type composite material by using a mixture of textile waste cuttings taken from the discarded trimmings of a garments producer and a textile manufacturer, cement and water. Their results indicated that textile off-cut waste can be used to prepare strong lightweight concrete with various possible uses in ceilings, walls, or as a wooden board substitute.

Agostini et al. [2007] investigated the use of dredged polluted sediment containing high concentrations of toxic elements such as Cd, Cu, Cr, Pb and Zn after treatment as partial and complete replacement of fine aggregate in cement mortar. Their results suggest that incorporation of sediment in mortar considerably increases the drying shrinkage of cement mortar. But incorporation of sediment in mortar does not affect water permeability and it significantly increases strength for low to moderate substitution levels, while the incorporation of large amounts of sediment leads to strength of the same order as that of the reference mortar.

7.5. Carpet Waste

Several references are available on the use of waste carpet fibre, mainly nylon and polypropylene, in concrete [Schmidt and Cieslak 2008, Vilkner et al. 2004, Wang 1999, Wang et al. 1994]. Concrete that contains carpet fibre up to 2% volume fraction exhibited lower compressive strength, comparable flexural strength, higher splitting tensile strength, higher toughness and lower drying shrinkage than conventional concrete. The compressive strength nonetheless meets the standard for normal and high strength concrete [Wang 1999, Wang et al. 1994]. Measurement of surface energy of concrete that contains carpet waste indicates the formation of a strong and water-resistant bond between carpet fibre and concrete [Schmidt and Cieslak 2008]. The incorporation of carpet fibre to improve mechanical and thermal properties of glass concrete is also reported [Vilkner et al. 2004].

7.6. Mining Waste

Yellishetty et al. [2008] reported that the strength level of concrete that contains iron-ore mineral waste of 12.5 mm and 20 mm size range as aggregate was better than that of concrete which includes conventional aggregate; this is because of an internal curing effect whereby water from the aggregate was gradually released into the concrete to further hydrate the paste. Leaching results indicated a very low level of dissolution of the toxic element into the leaching solution.

Colliery spoil (minestone), a by-product of coal mining, can be used as coarse and fine aggregate in the production of low and medium category concrete, usable as blinding concrete, and for use in bound granular fill or foundations. This material does have some drawbacks, such as excessive wear, expansive behaviour, leaching of toxic elements and radioactivity [Kinuthia et al. 2009].

Negm and Abouzeid [2008] reported that coarse solid phosphate tailings could be used as a coarse aggregate to prepare concrete mixes without seriously hindering the concrete's mechanical properties.

Conclusion

Significant progress has been made in the use of waste materials as aggregate, fibre or filler in concrete. The use of waste materials as aggregate in concrete consumes a vast amount of waste materials and saves natural resources. If waste materials are used in concrete as aggregate, fibre or filler this significantly changes its properties. The incorporation of some waste materials such as ground plastics, shredded rubber and coal bottom ash can enhance some specific properties of concrete and therefore special types of concrete can be developed by using these waste materials. The production of cheaper and more durable concrete by combing different mineral additions into cement and waste aggregate will go some way to helping to solve ecological and environmental problems.

Lack of widespread reliable data on aggregate substitutes can hinder the use of most waste materials as aggregate in concrete. More research is required to design consistent and durable concrete containing such waste aggregate. There is an increasing demand for a better understanding of concrete properties as well as for better control of the microstructure that develops in the concrete which incorporates various waste materials.

Acknowledgments

Nabajyoti Saikia acknowledges the FCT, Portugal for the financial assistance provided for this work.

References

ACAA (American Coal Ash Association) (2007). *Coal Combustion Product (CCP): Production and Use*. Aurora, American Coal Ash Association.

Aggarwal, P.; Aggarwal, Y.; Gupta, S. M. Effect of bottom ash as replacement of fine aggregates in concrete. *Asian Journal of Civil Engineering (Building and Housing) 2007,* 8, 49-62.

Agostini, F.; Skoczylas, F.; Lafhaj, Z. About a possible valorisation in cementitious materials of polluted sediments after treatment. *Cement Concrete Comp* 2007, 29, 270-278.

Ahmadi, B.; Al-Khaja, W. Utilization of paper waste sludge in the building construction industry. *Resour Conserv Recy* 2001, 32, 105–113.

Albano, C.; Camacho, N.; Hernandez, M.; Matheus, A.; Gutierrez, A. Influence of content and particle size of waste pet bottles on concrete behaviour at different w/c ratios. *Waste Manage 2009*, 29, 2707–2716.

Ali, N. A.; Amos, A. D.; Roberts M. 1993. Use of ground rubber tires in Portland cement concrete. In R.K. Dhir, (Ed.), *International Conference on Concrete 2000,* University of Dundee, Scotland, UK, 379–390.

Al-Jabri, K. S.; Hisada, M.; Al-Oraimi, S. K.; Al-Saidy, A. H. Copper slag as sand replacement for high performance concrete. Cement Concrete Comp 2006, 31, 483-488.

Al-Manaseer, A.A.; Dalal, T.R. Concrete containing plastic aggregates. *Concrete International* 1997, 19, 47–52.

Al-Negheimish, A. I.; Al-Sugair, F. H.; Al-Zaid, R. Z. Utilization of local steel making slag in concrete. *Journal of King Saud University - Engineering Sciences* 1997, 9, 39-55.

Al-Otaibi, S. Recycling steel mill scale as fine aggregate in cement mortars. *European Journal of Scientific* Research 2008, 24, 332-338.

Alter, H. The composition and environmental hazard of copper slags in the context of the Basel Convention. *Resour Conserv Recy* 2005, 43, 353-360.

Altun, I. A.; Yılmaz, I. Study on steel furnace slags with high MgO as additive in Portland cement. *Cement Concrete Res* 2002, 32, 1247–1249.

Akcaozoglu, S.; Atis, C. D.; Akcaozoglu, K. An investigation on the use of shredded waste PET bottles as aggregate in lightweight concrete. *Waste Manage* 2010, 32, 285-290.

Andrade, L. B.; Rocha, J. C.; Cheriaf, M. Evaluation of concrete incorporating bottom ash as a natural aggregates replacement. *Waste Manage* 2007, 27, 1190–1199.

Andrade, L. B.; Rocha, J. C.; Cheriaf, M. Influence of coal bottom ash as fine aggregate on fresh properties of concrete. *Constr Build Mater* 2009, 23, 609–614.

Asokan, P.; Osmani, M.; Price, A. D. F. Improvement of the mechanical properties of glass fibre reinforced plastic waste powder filled concrete. *Constr Build Mater* 2009, 24, 448-460.

Aspiras, F. F.; Manalo, J. R. I. Utilization of Textile Waste Cuttings as Building Material. *J Mater Process Technol* 1995, 48, 379-384.

Atzeni, C.; Massidda, L.; Sanna, U. Use of granulated slag from lead and zinc processing in concrete technology. *Cement Concrete Res* 1996, 26, 1381-1388.

Aubert, J. E.; Husson, B.; Vaquier, A. Use of municipal solid waste incineration fly ash in concrete. *Cement Concrete* Res 2004, 34, 957-963.

Ayano, T.; Sakata K. (2000). Durability of concrete with copper slag fine aggregate. *Fifth CANMET/ACI international conference on durability of concrete, SP-192,* American Concrete Institute, Farmington Hills, MI, 141–158.

Babcock and Wilcocx, 1978. *Steam - Its Generation and Use.* 39th Edition, Babcock &. Wilcox Co., New York, N.Y.

Baffa, I. P. D.; Akasaki J. (2005). *Light-concrete with leather: Durability aspects*, http://www.ppgec.feis.unesp.br/producao2005/Light-concrete%20with%20leather%20 durability%20aspects.pdf (online) (Accessed date: 15/01/2010).

Bai, Y.; Darcy, F.; Basheer, P. A. M. Strength and drying shrinkage properties of concrete containing furnace bottom ash as fine aggregate. *Constr Build Mater* 2005, 19, 691–697.

Basheer, P. A. M.; Bai, Y. Strength and durability of concrete with ash aggregate. Structures and Building 2005, 158, 191-199.

Basri, H. B.; Mannan, M. A.; Zain, M. F. M. Concrete using waste oil palm shells as aggregate. *Cement Concrete Res* 1999, 29, 619–622.

Batayneh; M., Marie; I., Asi, I. Use of selected waste materials in concrete mixes. Waste Manage 2007, 27, 1870–1876.

Bayasi, Z.; Zeng, J. Properties of polypropylene fiber reinforced concrete. *ACI Mater J* 1993, 90, 605–610.

Behnood, A. (2005). Effects of high temperatures on high-strength concrete incorporating copper slag aggregates. *Seventh international symposium on high-performance concrete*, SP-228-66, Washington, USA, 1063–1075.

Benazzouk, A.; Mezreb, K.; Doyen, G.; Goullieux, A.; Queneudec, M. Effect of rubber aggregates on the physico-mechanical behaviour of cement–rubber composites-influence of the alveolar texture of rubber aggregates. *Cement Concrete Comp* 2003, 25, 711–720.

Benazzouk, A.; Douzane, O.; Langlet, T.; Mezreb, K.; Roucoult, J. M.; Queneudec, M. Physico-mechanical properties and water absorption of cement composite containing shredded rubber waste. *Cement Concrete Comp* 2007, 29, 732–740.

Biswas A.K., Davenport W.G., *Extractive metallurgy of copper,* 1st edition, Pergamon Press OXFORD, UK, 1976.

Caliskan, S.; Behnood, A. Recycling copper slag as coarse aggregate: hardened properties of concrete, In *Seventh international conference on concrete technology in developing countries,* Malaysia, 2004; pp 91–98.

Caponero, J.; Tenorio, J. A. S. Laboratory testing of the use of phosphate-coating sludge in cement clinker. *Resour Conserv Recy* 2000, 29, 169–179.

Cheminos, J. M.; Fernandez, A. I.; Nadal, R.; Espiell, F. Short-term natural weathering of MSWI bottom ash. *J Hazard Mater* 2000, 79, 287–299.

Chen, M.; Zhou, M.; Wu, S. Optimization of blended mortars using steel slag sand. *J Wuhan Univ Tech* 2007, 22, 741-744.

Cheriaf, M.; Rocha, J. C.; Pera, J. Pozzolanic properties of pulverized coal combustion bottom ash. *Cement Concrete Res* 1999, 29, 1387–1391.

Choi, Y. W.; Moon, D. J.; Kim, Y. J.; Lachemi, M. Characteristics of mortar and concrete containing fine aggregate manufactured from recycled waste polyethylene terephthalate bottles. *Constr Build Mater* 2009, 23, 2829–2835.

Choi, Y. W.; Moon, D. J.; Chung, J. S.; Cho, S. K. Effects of waste PET bottles aggregate on the properties of concrete. *Cement Concrete Res* 2005, 35, 776–781.

Cinquepalmi, M. A.; Mangialardi, T.; Panei, L.; Paolini, A. E.; Piga L. Reuse of cement-solidified municipal incinerator fly ash in cement mortars: Physico-mechanical and leaching characteristics. *J Hazard Mater* 2008,151, 585–593.

Collins, R.J.; Ciesielski, S.K. *Recycling and use of waste materials and by-products in highway construction.* NCHRP (National Cooperative Highway Research Program, Synthesis of Highway Practice), Transportation Research Board, ISSN: 0547-5570, 500 Fifth Street, NW, Washington, DC 20001 USA, 1994; Issue No. 199.

Collivignarelli, C.; Sorlini, S.; Reuse of municipal solid waste incineration fly ashes in concrete mixtures. *Waste Manage* 2002, 22, 909–912.

Cyr, M.; Ludmann, C. Low risk meat and bone meal (MBM) bottom ash in mortars as sand replacement. *Cement Concrete Res* 2006, 36, 469 – 480.

Eldin, N. N.; Senouci, A. B. Rubber-tire particles as concrete aggregate. *J Matern Civil Eng* 1993, 5, 478-496.

Espinosa, D. C. R.; Tenorio, J. A. S. Laboratory study of galvanic sludge's influence on the clinkerization process. *Resour Conserv Recy* 2000, 31, 71–82.

Faraone, N.; Tonello, G.; Furlani, E.; Maschio, S. Steelmaking slag as aggregate for mortars: Effects of particle dimension on compression strength. *Chemosphere* 2009, 77, 1152–1156.

Fattuhi, N. I.; Clark, N. A. Cement-based materials containing tire rubber. *Constr Build Mater* 1996, 10, 229–236.

Galbenis, C. T.; Tsimas S. Use of construction and demolition waste as raw materials in cement clinker production. *China Particuology*. 2006, 4, 83-85.

Gallardo, R. S.; Adajar, M. A. Q. Structural performance of concrete with paper sludge as fine aggregates partial replacement enhanced with admixtures. *Symposium on Infrastructure Development and the Environment 2006,* 7-8 December 2006, SEAMEO-INNOTECH, University of the Philippines, Diliman, Quezon City, Philippines, 2006.

Ganjian, E.; Khorami, M.; Maghsoudi, A. A. Scrap-tyre-rubber replacement for aggregate and filler in concrete. *Constr Build Mater* 2009, 23, 1828–1836.

Ghafoori, N.; Bucholc, J. Investigation of lignite-based bottom ash for structural concrete. *J Mater Civil Eng* 1996, 8, 128-137.

Gines, O.; Chimenos, J. M.; Vizcarro, A.; Formosa, J.; Rosell, J. R. . Combined use of MSWI bottom ash and fly ash as aggregate in concrete formulation: Environmental and mechanical considerations. *J Hazard Mater* 2009, 169, 643–650.

Guneyisi, E.; Gesoglu, M.; Ozturan, T. Properties of rubberized concretes containing silica fume. *Cement Concrete Res* 2004, 34, 2309–2317.

Hernandez-Olivares, F.; Barluengab, G. Fire performance of recycled rubber-filled high-strength concrete. *Cement Concrete Res* 2004, 34, 109– 117.

Hwang, C. L.; Laiw, J. C. Properties of concrete using copper slag as a substitute for fine aggregate. *3rd international conference on fly ash, silica fume, slag, and natural pozzolans in concrete*, SP-114-82, American Concrete Institute, Farmington Hills, MI, 1989; pp 1677–1695.

Ismail, Z. Z.; Al-Hashmi, E. A. Use of waste plastic in concrete mixture as aggregate replacement. *Waste Manage* 2008, 28, 2041–2047.

Jumaat, M. Z.; Alengaram, U. J.; Mahmud, H. Shear strength of oil palm shell foamed concrete beams. *Mater Design* 2009, 30, 2227–2236.

Kan, A.; Demirboga, R. A novel material for lightweight concrete production. *Cement Concrete Comp 2009a*, 31, 489–495.

Kan, A.; Demirboga, R. A new technique of processing for waste-expanded polystyrene foams as aggregates. *J Mater Process Tech* 2009b, 209, 2994–3000.

Kang, J.; Han, C.; Zhang, Z. Strength and shrinkage behaviors of roller-compacted concrete with rubber additives. *Mater and Struct* 2009, 42, 1117–1124.

Kang, J.; Jiang, Y. Improvement of cracking-resistance and flexural behavior of cement-based materials by addition of rubber particles. *J Wuhan Univ Technol* 2008, 23, 579-583.

Kasemsaisiri, R.; Tantermsirikul, S. Properties of self-compacting concrete incorporating bottom ash as a partial replacement of fine aggregate. *ScienceAsia* 2008, 34, 87-95.

Khaloo, A. R.; Dehestani, M.; Rahmatabadi, P. Mechanical properties of concrete containing a high volume of tire–rubber particles. *Waste Manage* 2008, 28, 2472–2482.

Khanzadi, M.; Behnood, A. Mechanical properties of high-strength concrete incorporating copper slag as coarse aggregate. *Constr Build Mater* 2009, 23, 2183-2188.

Khatib, Z. K.; Bayomy, F. M. Rubberized portland cement concrete. *J Mater Civil Eng* 1999, 11, 206–213.

Khanbilvardi, R.; Afshari, S. Sludge ash as fine aggregate for concrete mix. *J Environ Eng-ASCE* 1999, 121, 633-638.

Kim, S. B.; Yi, N. H.; Kim, H. Y.; Kim, J. H. J.; Song, Y. C. Material and Structural Performance Evaluation of Recycled PET Fibre Reinforced Concrete. *Cement Concrete Comp* 2009, 32, 232-240.

Kinuthia, J.; Snelson, D.; Gailius, A. Sustainable medium-strength concrete (CS-concrete) from colliery spoil in South Wales UK. *Journal of Civil Engineering and Management* 2009, 15(2) 149-157.

Knuttgen and Muench, no date [online]. *Cement Production in Pavement Interactive.* Knuttgen, R & Muench S. (no date). The online community for all things pavement containing over 523 articles and 3,584 files, [on line], last modified: 28 May 2009 by Celeste Hoffman, Available from http://pavementinteractive.org/index.php?title= Cement_Production. [Accessed: 31st December 2009].

Kou, S. C.; Poon, C. S. *Properties of concrete prepared with crushed fine stone, furnace bottom ash and fine recycled aggregate as fine aggregates.* Constr Build Mater 2009b, 23, 2877–2886.

Kou, S. C.; Lee, G.; Poon, C. S.; Lai, W. L. *Properties of lightweight aggregate concrete prepared with PVC granules derived from scraped PVC pipes Waste Manage* 2009a, 29, 621–628.

Kumaran, G. S.; Mushule, N.; Lakshmipathy, M. A review on construction technologies that enables environmental protection: Rubberized concrete. *American Journal of Engineering and Applied Sciences* 2008, 1, 40-44.

Kuo, W. Y.; Huang, J. S.; Tan, T. E. Organo-modified reservoir sludge as fine aggregates in cement mortars. *Constr Build Mater* 2007, 21, 609–615.

Kurama, H.; Topcu, I. B.; Karakurt, C. Properties of the autoclaved aerated concrete produced from coal bottom ash. *J Mater Process Tech* 2009, 209, 767-773.

Lee, H. K.; Kim, H. K.; Hwang, E. A. Utilization of power plant bottom ash as aggregates in fibre-reinforced cellular concrete. *Waste Manage* 2010, 30, 274-284.

Lee, H. S.; Lee, H.; Moon, J. S.; Jung, H. W. Development of tire added latex concrete. *ACI Mater J* 1998, 95, 356–364.

Lee, M. H.; Lee, J. C. Study on the cause of pop-out defects on the concrete wall and repair method. *Constr Build Mater* 2009, 23, 482–490.

Li, F. Test research on copper slag concrete. *Journal of Fuzhou University* 1999, 127, 59–62.

Li, G.; Stubblefield, M. A.; Garrick, G.; Eggers, J.; Abadie, C.; Huang, B. Development of waste tire modified concrete. *Cement Concrete Res* 2004, 34, 2283-2289.

Lun, Y.; Zhou, M.; Cai, X.; Xu, F. Methods for improving volume stability of steel slag as fine aggregate. *J Wuhan Univ Tech* 2008, 23, 737-742.

Luxan, M. P.; Sotolongo, R.; Dorrego, F.; Herrero, E. Characteristics of the slags produced in the fusion of scrap steel by electric arc furnace. *Cement Concrete Res* 2000, 34, 517-519.

Mahieux, P. Y.; Aubert, J. E.; Escadeillas, G. Utilization of weathered basic oxygen furnace slag in the production of hydraulic road binders. *Constr Build Mater* 2009, 23, 742–747.

Mannan, M. A.; Alexander, J.; Ganapathy, C.; Teo, D. C. L. Quality improvement of oil palm shell (OPS) as coarse aggregate in lightweight concrete. *Build Environ* 2006, 41, 1239–1242.

Mannan, M. A.; Ganapathy, C.; Long-term strengths of concrete with oil palm shell as coarse aggregate. *Cement Concrete Res* 2001a, 31, 1319–1321.

Mannan, M. A.; Ganapathy, C.; 2001b. Mix design for oil palm shell concrete. *Cement and Concrete Research 2001b*, 31, 1323–1325.

Mannan, M. A.; Ganapathy, C. Engineering properties of concrete with oil palm shell as coarse aggregate. *Constr Build Mater* 2002, 16, 29-34.

Mannan, M. A.; Ganapathy, C. Concrete from an agricultural waste-oil palm shell (OPS). *Build Environ* 2004, 39, 441-448.

Manso, J. M.; Polanco, J. A.; Losanez, M.; Gonzalez, J. J. Durability of concrete made with EAF slag as aggregate. *Cement Concrete Comp* 2006, 28, 528–534.

Marzouk, O. Y.; Dheilly, R. M.; Queneudec, M. Valorisation of post-consumer waste plastic in cementitious concrete composites. *Waste Manage* 2007, 27, 310–318.

Maslehuddin, M.; Sharif, A. M.; Shameem, M.; Ibrahim, M.; Barry, M. S. Comparison of properties of steel slag and crushed limestone aggregate concretes. *Constr Build Mater* 2003, 17, 105–112.

Metwally, M. E. A.;, Seleem, M. H., Balaha, M. M.; Abd El-Rahman, H. Utilizing of slag produced from recycling of spent lead-batteries as concrete aggregate. *Alexandria Engineering Journal* 2005, 44, 883-892.

Monosi, S.; Giretti, P.; Moriconi, G.; Favoni, O.; Collepardi, M. *Non ferrous slag as cementitious material and fine aggregate for concrete.* Special Publication, American Concrete Institute, 2001; Vol. 202, 33-44.

Monshi, A.; Asgarani M. K. Producing Portland cement from iron and steel slags and limestone. *Cement Concrete Res* 1999, 29, 1373–1377.

Morrison, C.; Hooper, R.; Lardner, K. The use of ferro-silicate slag from ISF zinc production as a sand replacement in concrete. *Cement Concrete Res* 2003, 33, 2085–2089.

Morrison, C.; Richardson, D. Re-use of zinc smelting furnace slag in concrete. *Engineering Sustainability* 2004,157, 213 –218.

Moura, W.; Masuero, A.; Dal Molin, D.; Vilela A. Concrete performance with admixtures of electrical steel slag and copper concerning mechanical properties. *2nd CANMET/ACI international conference on high-performance concrete*, SP-186 American Concrete Institute, Farmington Hills, MI, 1999; 81–100.

Naik, T. R. Greener concrete using recycled materials. *Concrete International* 2002, 24, 49-53.

Naik, T. R.; Friberg, T. S., Chun, Y. M. Use of pulp and paper mill residual solids in production of cellucrete. *Cement Concrete Res* 2004, 34, 1229–1234.

Naik, T. R.; Singh, S. S.; Huber, C. O.; Brodersen, B. S. Use of post-consumer waste plastics in cement-based composites. *Cement Concrete Res* 1996, 26, 1489-1492.

Negm, A. A.; Abouzeid, A. Z. M. Utilization of solid waste from phosphate processing plants. *Physicochemical Problems of Mineral Processing* 2008, 42, 5-16.

Ozkan, O.; Yuksel, I.; Muratoglu, O. Strength properties of concrete incorporating coal bottom ash and granulated blast furnace slag. *Waste Manage* 2007, 27, 161–167.

Ozturk, T.; Bayraklı, M. (2005). The possibilities of using tobacco waste in producing lightweight concrete. *Agricultural Engineering International: the CIGR Ejournal*. Vol. VII, August, Manuscript BC 05 006.

Pang, X.; Liu, C.; Suri, R. Recycling spent mushroom substrate as fine aggregates in concrete. *TRB 86th Annual Meeting Compendium of Papers CD-ROM, Transportation Research Board 86th Annual Meeting, Transportation Research Board,* 500 Fifth Street, NW, Washington, DC 20001 USA, Paper #07-1726; 2007.

Panyakapo, P.; Panyakapo, M. Reuse of thermosetting plastic waste for lightweight concrete. *Waste Manage* 2008, 28(9) 1581–1588.

Park, S. B.; Jang, Y. I.; Lee, J.; Lee, B. J. An experimental study on the hazard assessment and mechanical properties of porous concrete utilizing coal bottom ash coarse aggregate in Korea. *J Hazard Mater* 2009, 166, 348–355.

Pellegrino, C.; Gaddo, V. Mechanical and durability characteristics of concrete containing EAF slag as aggregate. *Cement Concrete Comp* 2009, 31, 663–671.

Penpolcharoen, M. Utilization of secondary lead slag as construction material. *Cement Concrete Res* 2005, 35, 1050-1055.

Pera, J.; Coutaz, L.; Ambroise, J.; Chababbet, M. Use of incinerator bottom ash in concrete. *Cement Concrete Res* 1997, 27(1) 1-5.

Pereira, D. A.; de Aguiar, B.; Castro, F.; Almeida, M. F.; Labrincha, J. A. Mechanical behaviour of Portland cement mortars with incorporation of Al-containing salt slags. *Cement Concrete Res* 2000, 30, 1131-1138.

Pierce, C. E.; Blackwell, M. C. Potential of scrap-tire rubber as lightweight aggregate in flowable fill). *Waste Manage* 2003, 23, 197-208.

Proctor, D. M.; Fehling, K. A.; Shay, E. C.; Wittenborn, J. L.; Green, J. J.; Avent, C.; Bigham, R. D.; Connolly, M.; Lee, B.; Shepker, T. O.; Zak, M. A. Physical and chemical characteristics of blast furnace, basic oxygen furnace, and electric arc furnace steel industry slags. *Environ Sci Technol* 2000, 34, 1576-1582.

Qasrawi, H.; Shalabi, F.; Asi, I. Use of low CaO unprocessed steel slag in concrete as fine aggregate. *Constr Build Mater* 2009, 23, 1118–1125.

Qian, G.; Sun, D. D.; Tay, J. H.; Lai, Z.; Xu, G. Autoclave properties of kirschsteinite-based steel slag. *Cement Concrete Res* 2002, 32, 1377–1382.

Raghvan, D.; Huynh, H.; Ferraris, C. F. Workability, mechanical properties and chemical stability of a recycled tire rubber-filled cementitious composite. *J Mater Sci* 1998, 33, 1745–1752.

Rao, S. M.; Venkatarama Reddy, B. V.; Lakshmikanth, S.; Ambika, N. S. Re-use of fluoride contaminated bone char sludge in concrete. *J Hazard Mater* 2009, 166, 751–756.

Ravina, D. Properties of fresh concrete incorporating a high volume of fly ash as partial fine sand replacement. *Mater Struct* 1997, 30, 473–479.

Reddy, A. S.; Pradhan, R. K.; Chandra, S. Utilization of Basic Oxygen Furnace (BOF) slag in the production of a hydraulic cement binder. *Int J Miner Process* 2006, 79, 98-105.

Rodriguez, A.; Manso, J. M.; Aragon, A.;, Gonzalez, J. J. Strength and workability of masonry mortars manufactured with ladle furnace slag. *Resour Conserv Recy* 2009, 53, 645–651.

Rogbeck, J.; Knutz, A. Coal bottom ash as light fill material in construction. *Waste Manage* 1999, 16, 125-128.

Rojas, M. F.; Sanchez de Rojas, M. I. Chemical assessment of the electric arc furnace slag as construction material: Expansive compounds. *Cement Concrete Res* 2004, 34, 1881–1888.

Rostami, H.; Lepore, J.; Silverstraim, T.; Zundi, I. (1993). Use of recycled rubber tires in concrete. In: Dhir R.K. (Ed.), *International Conference on Concrete 2000*, University of Dundee, Scotland, UK, 391–399.

Saikia, N.; Kato, S.; Kojima, T. Production of cement clinkers from municipal solid waste incineration (MSWI) fly ash. *Waste Manage* 2007, 27, 1178–1189.

Saikia, N.; Cornelis, G.; Mertens, G.; Elsen, J.; Van Balen, K.; Van Gerven, T.; Vandecasteele, C. Assessment of Pb-slag, MSWI bottom ash and boiler and fly ash for using as a fine aggregate in cement mortar. *J Hazard Mater* 2008, 144, 766-777.

Sales, A.; De Souza, F. R. Concretes and mortars recycled with water treatment sludge and construction and demolition rubble. *Constr Build Mater* 2009, 23, 2362–2370.

Schmidt, H.; Cieslak, M. Concrete with carpet recyclates: Suitability assessment by surface energy evaluation. *Waste Manage.* 2008, 28, 1182-1187.

Segre, N.; Joekes, I. Use of tire rubber particles as addition to cement paste. *Cement Concrete Res* 2000, 30, 1421-1425.

Setien, J.; Hernandez, D.; Gonzalez, J. J. Characterization of ladle furnace basic slag for use as a construction material. *Constr Build Mater* 2009, 23, 1788–1794.

Shi, C. Characteristics and cementitious properties of ladle slag fines from steel production. *Cement Concrete Res* 2002, 32, 459–462.

Shi, C.; Qian, J. High performance cementing materials from industrial slag - a review. *Resour Conserv Recy* 2000, 29, 195–207.

Shi, C.; Meyer, C.; Behnood, A. Utilization of copper slag in cement and concrete. *Resour Conserv Recy 2008*, 52, 1115-1120.

Shoya, M.; Nagataki, S.; Tomosawa, F.; Sugita, S.; Tsukinaga Y. Freezing and thawing resistance of concrete with excessive bleeding and its improvement. *Fourth CANMET/ACI international conference on durability of concrete,* SP-170-45 American Concrete Institute, Farmington Hills, MI, 1997; pp 879–898.

Siddique, R. Effect of fine aggregate replacement with Class F fly ash on the mechanical properties of concrete. *Cement Concrete Res* 2003a, 33, 539–547.

Siddique R., 2003b. Effect of fine aggregate replacement with Class F fly ash on the abrasion resistance of concrete. *Cement Concrete Res* 2003b, 33, 1877–1881.

Siddique, R.; Naik, T. R. Properties of concrete containing scrap-tire rubber - an overview. *Waste Manage* 2004, 24, 563-569.

Siddique, R.; Khatib, J.; Kaur, I. *Use of recycled plastic in concrete: A review. Waste Manage* 2008, 28, 1835–1852.

Silva, D. A.; Betioli, A. M.; Gleize, P. J. P.; Roman, H. R.; Gomez, L. A.; Ribeiro, J. L. D. Degradation of recycled PET fibres in Portland cement-based materials. *Cement Concrete Res* 2005, 35, 1741-1746.

Sorlini, S.; Collivignarelli, C.; Plizzari, G.; Foglie, M. D. Reuse of Waelz slag as recycled aggregate for structural concrete. In *International RILEM Conference on the Use of Recycled Materials in Building and Structures;* Vazquez, E.; Hendriks Ch. F. & Janssen, G. M. T. (Eds); ISBN: 2-912143-52-7, e-ISBN: 2912143756; 2004; 1086-1094.

Soroushian, P.; Plasencia, J. S.; Ravanbakhsh, S.; Assessment of reinforcing effects of recycled plastic and paper in concrete. *ACI Mater J.* 2003, 100, 203–207.

Sukontasukkul, P.; Chaikaew, C. Properties of concrete pedestrian block mixed with crumb rubber. *Constr Build Mater* 2006, 20, 450–457.

Taeb, A.; Faghihi, S. Utilization of copper slag in the cement industry. *ZKG Int* 2002, 55, 98–100.

Tandon, V.; Picornell, M. (1998). *Safe disposal of fly ash in pavement or earth structures not requiring high strength materials.* Geo-Congress, Oct 18-21 1998, Boston, MA, USA: ASCE, Reston, VA.

Taylor, H. F. W. *Cement Chemistry; 2nd Edition*; Thomas Telford Publishing: London, UK, 1997.

Teo, D.; Mannan, M. A.; Kurian, J. V. Flexural Behaviour of Reinforced Lightweight Concrete Beams Made with Oil Palm Shell (OPS). *Journal of Advanced Concrete Technology* 2006, 4, 459-468.

Teo, D. C. L.; Mannan, M. A.; Kurian, V. J.; Ganapathy C. Lightweight concrete made from oil palm shell (OPS): Structural bond and durability properties. *Build Environ* 2007, 42, 2614–2621.

Teo, D. C. L.; Mannan, M. A.; Kurian, V. Durability of lightweight OPS concrete under different curing conditions. *J. Mater Struct* 2010, 43, 1–13.

Topcu, I. B. The properties of rubberized concrete. *Cement Concrete Res* 1995, 25, 304–310.

Tossavainen, M.; Engstrom, F.; Yang, Q.; Menad, N.; Larsson, M. L., Bjorkman, B. Characteristics of steel slag under different cooling conditions. *Waste Manage* 2007, 27, 1335–1344.

Trezza, M. A.; Scian, A. N. Waste with chrome in the Portland cement clinker production. *J Hazard Mater* 2007, 147, 188–196.

Turatsinze, A.; Bonnet, S.; Granju, J. -L. Potential of rubber aggregates to modify properties of cement based-mortars: Improvement in cracking shrinkage resistance. *Constr Build Mater* 2007, 21, 176–181.

Vangelatos, I.; Angelopoulos, G. N.; Boufounos, D. Utilization of ferroalumina as raw material in the production of Ordinary Portland Cement. *J Hazard Mater* 2009, 168, 473–478.

Vilkner, G.; Meyer, C.; Shimanovich, S. Properties of glass concrete containing recycled carpet fibers. In *6th International RILEM Symposium on Fibre Reinforced Concrete, Varenna*, Italy, Sept. 20-22, 2004, 1431 - 1440.

Wang, Y. Utilization of recycled carpet waste fibers for reinforcement of concrete and soil. *Polymer-Plastics Polym-Plast Technol* 1999, 38, 533-546.

Wang, Y.; Zureick, A. H.; Cho, B. S.; Scott D. E. Properties of fibre reinforced concrete using recycled fibres from carpet industrial waste. *J Mater Sci* 1994, 29, 4191-4199.

Wilson, A. (1993). Cement and concrete: Environmental considerations. Environmental Building News [on line], 2 (2). Available from http://www.buildinggreen.com/auth/article.cfm/1993/3/1/Cement-and-Concrete-Environmental-Considerations/ [Accessed: 31st December 2009].

Xue, Y.; Wu ,S.; Hou, H.; Zha, J. Experimental investigation of basic oxygen furnace slag used as aggregate in asphalt mixture. *J Hazard Mater* 2006, 138, 261–268.

Yang, E. I.; Yi, S. T.; Leem, Y. M. Effect of oyster shell substituted for fine aggregate on concrete characteristics: Part I. Fundamental properties. *Cement Concrete Res* 2005, 35, 2175-2182.

Yang, E. I.; Kim, M. Y.; Park, H. G.; Yi, S. T. Effect of partial replacement of sand with dry oyster shell on the long-term performance of concrete. *Constr Build Mater* 2009, (in press, doi: 10.1016/j.conbuildmat. 2009.10.032).

Yellishetty, M.; Karpe, V.; Reddy, E. H.;, Subhash, K. N.;, Ranjith, P. G. Reuse of iron ore mineral waste in civil engineering constructions: A case study. *Resour Conserv Recy* 2008, 52, 1283–1289.

Yesilata, B.; Isıker, Y.; Turgut, P. Thermal insulation enhancement in concretes by adding waste PET and rubber pieces. *Constr Build Mater* 2009, 23, 1878–1882.

Yilmaz, A.; Degirmenci, N. Possibility of using waste tire rubber and fly ash with Portland cement as construction materials. *Waste Manage* 2009, 29, 1541–1546.

Yuksel, I.; Bilir, T. Usage of industrial by-products to produce plain concrete elements. *Constr Build Mater 2007*, 21, 686–694.

Zelic, J. Properties of concrete pavements prepared with ferrochromium slag as concrete aggregate. *Cement Concrete Res* 2005, 35, 2340-2349.

Zong, L. The replacement of granulated copper slag for sand concrete. *Journal of Qingdao Institute of Architecture and Engineering,* 2003, 24, 20–22.

INDEX

D

E

F

M

N

O

P

Q

R

S

T

U

V

W

X

Y

Z